ROBOTIZATION AND ECONOMIC DEVELOPMENT

This book critically examines the sweeping impacts of robotization and the use of artificial intelligence on employment, per capita income, quality of life, poverty, and inequality in developing and developed economies.

It analyzes the direct and indirect effects they have had and are projected to have on the labour markets and production processes in the manufacturing, healthcare, and agricultural industries, among others. The author explores comparisons of human labour with robotic labour emphasizing the changes that new technologies will bring to traditionally labour-intensive industries. Offering various insights into the effectiveness, benefits, and negative implications of robotization on the economy, the book provides a comprehensive picture for policymakers to implement changes that embrace new technologies while meeting employment needs and development goals.

Topical and lucid, this book will be an essential read for scholars and researchers of science and technology studies, digital humanities, economics, labour studies, public policy, development studies, political studies, and sociology as well as policymakers and others who are interested in these areas.

Siddhartha Mitra is Professor of Economics, Jadavpur University, Kolkata, West Bengal, India.

'Such a fascinating look at the large, even humongous shifts that are all around us through the lens of history, technology, and economics. Whether you are a student or a professional who wishes to understand how robotization and AI are going to impact our world—this book is for you. Mitra takes a complex subject and connects the dots in ways that are unexpected, surprising, and ultimately very satisfying.'

—**Dr. Shalini Lal**, *Future of Work Specialist, UCLA PhD, and Fellow, Wharton, USA*

'The robot revolution is already underway, with profound implications for developing country and developed economies. In this wide-ranging examination, Siddhartha Mitra explains how and when new technologies will replace human labour in manufacturing, agriculture, and service provision. Employing basic economic tools and a clear depiction of how robots, AI, and machine learning will substitute for human effort, Mitra identifies which workers are most at risk of displacement and income loss. With Covid-19 accelerating the adoption of labour-replacing technology, it is urgent that we struggle with Mitra's recommendations for policies designed to harness the benefits of robotization while ensuring a just transition to the brave, new world on our doorstep.'

—**Mary E. Lovely**, *Professor, Maxwell School of Citizenship and Public Affairs and Senior Fellow, Peterson Institute, USA*

'Robotization, like other forms of technical progress, expands opportunities for the human race. But in a market economy, it also poses a threat by displacing human labour by robots. These aspects of robotization have been debated in recent times. This book by Professor Siddhartha Mitra provides an eminently readable discussion of the issues involved. He provides a historical background, and an analytical apparatus to make sense of the trade-offs involved. The book does not shy away from discussing difficult questions but does so in a lucid manner. I have no doubt, given the contents of the book, that it will be widely read. And the reader will appreciate, as I have, the hard work that went into making these seemingly obscure topics accessible'.

—**Partha Sen**, *Professor and Former Director, Delhi School of Economics, India*

ROBOTIZATION AND ECONOMIC DEVELOPMENT

Siddhartha Mitra

Routledge
Taylor & Francis Group

LONDON AND NEW YORK

Cover image: Getty Images

First published 2023
by Routledge
4 Park Square, Milton Park, Abingdon, Oxon OX14 4RN

and by Routledge
605 Third Avenue, New York, NY 10158

Routledge is an imprint of the Taylor & Francis Group, an informa business

British Library Cataloguing-in-Publication Data
A catalogue record for this book is available from the British Library

Library of Congress Cataloging-in-Publication Data
A catalog record has been requested for this book

ISBN: 978-0-367-35322-3 (hbk)
ISBN: 978-0-367-35609-5 (pbk)
ISBN: 978-0-429-34061-1 (ebk)

DOI: 10.4324/9780429340611

Typeset in Bembo
by Deanta Global Publishing Services, Chennai, India

CONTENTS

PREFACE AND ACKNOWLEDGEMENTS

My interest in robots evolved in 2017, when I came across writings in the press and was fortunate enough to read Martin Ford's *Rise of the Robots*. At that point, some of it seemed like science fiction to me, though constant exposure to similar material over the past four years has helped to dispel much of the awe.

Ford's documentation of the material was intensive as well as extensive and brilliant. I have always believed that interdisciplinary work, involving academics/researchers in a particular discipline learning from other disciplines, is much more geared to problem-solving in the real world than work that never crosses the confines of a discipline. Therefore, I am of the firm belief that the wealth of material on robots and artificial intelligence generated by Ford and other futurists could greatly benefit the construction of frameworks that rely on standard economic concepts and tools to generate predictions and prescriptions in regard to the impact of the AI revolution on unemployment, economic growth, poverty, and income equality.

Going back to 2017, I happened to exchange thoughts on this topic with Mousumi Das, another economist. Both of us were quite excited about the potential for research on the mentioned topic, though within the discipline, we could not see much of an eagerness for broad research. The models that we saw were based on very specific assumptions which lent themselves to precise mathematical treatment, but I was sceptical about how these could be used for framing economic policy that could enlarge the benefits and contain the downside of the AI revolution. I do not want my fellow economists to get me wrong in this regard: I am not against mathematical models based on specific assumptions and feel that these do very well in the generation of clear insights into important causal chains underlying important economic phenomena. But complexity is often traded off against clarity in these models and policymaking, which does not embrace the complexity of the real world in order to make it better, is hardly ideal.

Therefore, what we, as co-authors, started out to do was to generate a broad framework to answer the questions we set ourselves. Such a framework was necessarily open in nature and therefore we needed to eschew mathematical treatment of the subject. By doing so we also opened up the possibility of academic exchange with scholars in other disciplines, believers as we were of the problem-solving power of interdisciplinary research and study.

The work generated a couple of papers and their more concise and readable versions which appeared in a leading daily and a well-known portal. The content of these contributions has been included in the book, both explicitly and in an indirect manner. The papers received useful comments from Shalini Sehgal, Anirban Mitra, Devashish Mitra, and Ashok Kotwal. I had very enlightening conversations with Partha Sen and Jaideep Roy. As is usually the case, I was constantly learning more about the various facets of AI including recent developments such as machine learning. It therefore dawned on me that a paper was inadequate for answering the broad question that had been posed and a book was needed. The book, I argued with myself, would again be for the most part non-mathematical, and written in a style that would help me reach out to a wide audience and impress upon them the significance of the policy issues that were at stake. This is a considerable challenge: the study of AI is riddled with terms and so is the language that economists use. It was necessary for the book to carry explanations of all these concepts.

I approached Routledge with a book proposal and was very honoured that the proposal was accepted. The book has taken more than a year to write. Often, I would engage myself in writing for a month but would find myself diverted to other academic responsibilities: teaching, grading, work on academic papers with co-authors that could not be put off, and supervising young researchers. But I managed to work around those and get past the finishing line. I hope that this tiny book, which can be perused in quick time, serves as an adequate window for the economic issues linked to development of artificial intelligence and robots.

I am not going to discuss the contents of the book here as that discussion is part of the initial chapter. I would just end by mentioning a number of people who contributed to or catalyzed the writing of this book: Mousumi Das as the co-author of initial works that have fed into this book; those named above who have read these initial works and offered very useful comments; Vanshika Agarwal who has read the proofs for the book and helped in editing and formatting these; and Mansi Kedia who offered very useful comments on the entire draft of the book.

It deserves to be noted that there are authors other than those cited in the book who have shaped my foundations in the studied subject. I thank all the unmentioned authors and regret that I have not been able to include their names.

Last and definitely not the least, I would like to thank my mother, Rita Mitra, who had to endure my shutting off myself for hours in the house as I worked on the book during these Covid-ridden times. I dedicate this book to her.

Siddhartha Mitra
Kolkata, India

1
A QUICK SUMMARY FOR THE READER OF FORTHCOMING CHAPTERS

As is often the case in a book, this book covers a large number of topics and goes off at tangents to engage in discussions which are relevant to its mission. Though the layout of the book is presented crisply in the next chapter, it is possible that in going through detailed discussions and descriptions, a significant proportion of the readers will lose the woods for the trees. This chapter therefore attempts to provide a bird's eye view of the discussions in the forthcoming chapters by presenting these compactly in the following sections, an endeavour which enables the readers to see how the entire material hangs together.

Motivation, mission, concepts, and coverage

The author of any non-fiction book should ask herself the following questions:

- Why am I writing this book? How is it going to be useful? (Motivation)
- What am I hoping to accomplish through this book? (Mission)
- As fellow travellers on this mission, how am I going to communicate with the readers? Given that we are going to delve deep into one particular aspect of life and its implications for human happiness and behaviour, it is essential to explain the relevant terminology and establish a language of which we have a common understanding, irrespective of our past backgrounds. (Concepts)

The above is what the second chapter in this book tries to do. The following sub-sections deal with the subject of each of these questions.

DOI: 10.4324/9780429340611-1

Motivation for the book

Human workers have always been wary of machines that could do part or most of their work and deprive them of their jobs. This suspicion, and consequent fear, was manifested in the Luddites – weavers and textile workers in the late 18th century – breaking the machines that were displacing them from their jobs. This wariness exists to this very day. Polls point to significant proportions of the workforce feeling threatened by technology. History, however, indicates that the mass mechanization of production in the industrial revolution of the 18th and 19th centuries did not result in an increase in unemployment in spite of an increase in output per worker, the reason being the widening of the consumption base for existing goods and a vast increase in product variety heralded by the discovery of new goods such as the television and the motor car, which opened up entrepreneurial opportunities and new vistas for employment.

Could it be that history will repeat itself for machines in the artificial intelligence (AI) age? Critics, however, point out that the AI revolution is different from the industrial and agricultural revolutions of the past – it is powered by robots who can function without human interference for long periods of time and can to a large extent program themselves. Thus, the complementarity between man and machine which characterized the revolutions of the past is often absent when robots take over production in different sectors of the economy. This has adverse implications for the magnitude of human employment. However, the production process in many sectors of the economy is marked by uncertainty in the associated environment. Robots to date are not as good as humans at coping with these uncertainties. High incidence and wide range mark the uncertainties in the services sectors, which arise from the fact that in many of these, such as hospitality and entertainment, the provision of the product is based on the server (human or robot) dealing with human customers, whose demands, actions, and reactions exhibit a large variation. In comparison to humans, robots are deficient in soft and fuzzy skills, associated with the use of empathy and other feelings and emotions, needed in such sectors.

The mission of the book

The mission of this book was to speculate about the impact that the new technology, powered by AI, will have on employment, incidence of poverty, economic growth, and quality of life of the common man. Speculation was not as much about generating numerical estimates of unemployment, poverty, and growth of per capita income in the future as it was about identifying the various tendencies being unleashed by AI which could impact the future trajectories of mentioned variables. This is because the book recognizes that our future in this AI-driven world would depend significantly on the policies that the nations of the world implement. Quite naturally therefore the book comes up with suggestions for policies which, given the mentioned tendencies, can lead the world towards a bright future by leveraging the benefits of AI and containing its adverse impacts.

Given that robots are fundamentally different from traditional machines, going about the mentioned mission involved not only the use of historical data on how automation has impacted unemployment but its supplementation by information on recent developments in the field of robotics and how these have affected production processes as well as cost and efficiency of production.

Coverage and concepts

The book has tried to be very comprehensive in its coverage. As indicated, it has looked at various ways in which AI has been adopted by entrepreneurs in different sectors of the economy to transform production processes and the role of human labour. Two major influences in regard to the adoption of AI have been the advances in machine learning (ML) and the sudden onset of the Covid pandemic. Both influences and the body of theoretical and empirical work on the economic implications of robotization have been discussed at great length in the book to speculate about the future and arrive at policy recommendations.

The discussion in the book is based on the accurate understanding of a few key concepts: artificial intelligence as the human-like intelligence displayed by an inanimate object plus other types of intelligence displayed by inanimate objects such as the ability to capture images; *robot* as a receptacle of artificial intelligence; and finally *robotization*, which has been used to either refer to the introduction of robots in the production process, or the use of artificial intelligence in production, which involves one or more out of robots, computers, internet, smartphones, and the like.

The economic impact of robotization at the level of the producer and basic implications

Broadly, the expected impact of robotization in a sector is to decrease the amount of labour used in the production process through a fall in labour intensity of output. However, there are channels through which it can contribute to employment: an increase in output resulting from cost reductions, which tends to counteract the effect of reduced labour intensity in production and enhances use of labour in marketing and transportation of output. Moreover, the mentioned cost reduction also leads to a price fall; the resulting increase in purchasing power manifests itself in an increase in consumption and production across the board, another way in which employment is stimulated.

Given the above rudimentary sketch of an important aspect of the economic impact of robotization, the actual net impact varies across sectors in an economy according to the ease with which innovation can unearth a cost-efficient robotized technology.

Robotization and agriculture

In agriculture, the already low level of employment in developed countries is being reduced further as AI-based technology is being used to perform nuanced activities, such as picking fruits and flowers and pruning vines, until now the preserve of humans. In developing countries, the use of robots is much lower than that in developed countries, a result of the smaller size of farm holdings and poor purchasing power of farmers. However, this might not continue to be the case, given recent policy changes: for example, in India, recent legislative and policy changes promote consolidation of agricultural holdings, a development which would in turn spur robotization.

The incorporation of AI in agricultural practices has resulted in what is known as precision agriculture[1]: targeting plants precisely with the needed amounts of fertilizers, pesticides, water, etc., with plant health and needs determined separately for each plant through the collection of data. Given a reduction in the use of material inputs, a result of the mentioned precision, and the displacement of human labour engaged in many tasks by more efficient robots, tendencies are generated for a reduction in the overall cost of cultivation. At the same time, the mentioned precision in monitoring plant health and optimizing plant growth by catering to the specific input needs of each plant also enhances crop yield. Lower cost of cultivation and enhanced yields resulting from robotization imply that a switch to the robotized technology by farmers is a rational step; moreover, this step would cater to the nutritional needs of a growing global population characterized by significant growth in per capita income.

Robotization and manufacturing

Manufacturing has seen increased robotization all over the globe, an outcome of robots and computers becoming more efficient, powerful, and cheaper over time, and greater availability of information in the public domain about the relative advantages of robots in many types of production.

Robotization might not only result in a lower production cost per unit of output but there are other advantages of robots in comparison to humans: significantly lower training costs as programming of an army of robots to do a specific task does not involve the handholding needed to cater to the heterogeneity of human workers, and a greater ease of management given that human workers can engage in unrest and suffer from problems of mental health, especially when escalations in demand induce employers to introduce longer work shifts.

The rate of growth of employment of industrial robots has vastly exceeded the rate of growth of human employment in the recent past. If we assume that both types of employment cater to the same need in regard to completion of tasks, then the mentioned disparity in the rates of growth would translate to robotization catering to an ever-increasing proportion of the need and a reduction in growth of human employment. There is, however, the possibility that as costs of production go down with robotization, the resulting enhancement of output would give employment of human workers an upward push, which would counteract the direct displacement of human workers by robots in certain tasks.

A consequence which should be viewed with some concern is the tendency generated by robotization for developed countries to switch to insourcing from outsourcing: note that outsourcing was a reaction to labour costs in developed countries being much higher than that in developing countries; with this cost advantage being rendered redundant by cost-efficient robotized technologies becoming available in certain types of production it makes sense for developed countries to locate factories closer to the consumption hubs within their borders.

Robotization and services

In regard to services, robots have been employed across the board in the retail sector, restaurants and hotels, and hospitals as well as journalists, painters, and music composers to name a few professions. While the direct impact of robotization on human employment is negative, jobs which require complex communication with heterogeneous human clients and are based on interaction laced with emotions such as empathy are, as of now, immune from being taken over by robots. Finally, the use of AI implies that work hours will get reduced and more time would become available for leisure. This bodes well for the expansion of the hospitality and entertainment industries. Such expansion will add significantly to the magnitude of human employment as these industries provide services based on complex communication and the expression of emotion and warmth.

A special mention needs to be made of online courses and training programmes: it would be possible for a reputed instructor/institution to reach out to thousands of trainees/students all over the world, provide them with human capital, and certify the formation of this human capital through the award of degrees/diplomas. Thus, the cost of formal programmes for human capital formation would decline.

Robots would also serve as companions, not substitutes, to humans in providing geriatric care. This would happen because robots in this sector are geared to making the elderly more cheerful and active and ready to engage in a more varied life. This engagement would be associated with many younger humans getting employed and performing tasks related to lifting, cooking, fetching, engaging in detailed communication with the elderly, etc. The future might see schools for the elderly being talked about in the same vein all over the globe as schools for children, with the elderly being able to keep up with technological changes, etc. as a result of time spent in such schools.

At the frontiers of robotization: machine learning

The use of machine learning has greatly enhanced the benefits of automation. ML involves the use of data about human actions in regard to completion of tasks

– recognition of images, use of languages, and limb movements, among others – by computerized systems, which enables these systems to perform these tasks with greater speed and comparable accuracy as well as lower cost as compared to humans. Thus, these help to lower product prices and boost profits as well as enhance the scale of output of existing products. Creation of new products is also made possible, and this includes transformation of blueprints for products which were economically unviable earlier into actual items of consumption. Note that many of these tasks such as image and face recognition are accomplished by humans without knowing the underlying stepwise procedures and therefore humans themselves cannot write programs containing these steps; an ML system has the capability to exactly match the appropriate human response to each data input if it is trained on data regarding inputs and outputs.

Translation services, which were earlier rendered through the efforts of humans, are now being provided at a much lower cost and a higher scale by computerized systems. Another example is face recognition, which was earlier the preserve of humans: ML has now enabled computerized systems to compare an unlabelled human face with labelled images from a large catalogue and provide the basis for the maintenance of security and prevention of unauthorized entry in places such as libraries, train stations, airports, etc. Subscriptions in regard to movies and music by way of which offerings to millions of customers are made responsive to the preferences revealed by these very customers through their past choices (Netflix, YouTube, etc.) are products of the AI age, given that the requirement for processing of data involved in the provision of such services is immense and beyond the reach of human cognition. It is now possible for people to identify others – irrespective of where they are physically located – with similar tastes, inclinations, thoughts, and ideologies through appropriate software and search engines providing specific services (Twitter, Facebook, and Instagram) and form tightly knit communities.

A computerized system taking over human tasks such as translation has a direct negative impact on employment. But we can only imagine the leap in the commercial provision of translation services that such computerization has facilitated. This obviously has some positive implications for the employment of those who run enterprises for the provision of translation services; moreover, one can also imagine the boost the mentioned leap in the provision of translation services has provided, by breaking down linguistic barriers, in regard to online purchases of products and services. There is obviously a positive employment effect here as well as another created by the invention of new products mentioned in the last paragraph.

Finally, one must remember that many of these new products are provided to customers for free or at a nominal charge and these replace older ways of spending one's leisure time based on the expenditure of money, time, and fuel: for example, spending a lot of time in driving several kilometres to meet a friend. Social media such as Facebook, WhatsApp, and Instagram with billions of users enable social interaction without time- and fuel-intensive travel; the time and money saved can be used to consume goods and services, thus boosting the market demand for these. This leads to generation of incomes and employment.

The sudden exogenous shock to robotization provided by Covid

The book has mostly focused on how the logic of evolution in regard to capitalism and automation has enhanced the role of robots and AI in our lives and affected the prospects for economic development and human employment. While evolution proceeds on the basis of systematic tendencies, there are some influences that are sudden and completely unanticipated. These can be viewed as exogenous shocks.

One such massive exogenous shock is the Covid pandemic. In normal times, employers can be reluctant about terminating a very large number of jobs and undertaking matching robotization because of the demoralizing effect that such termination can have on those remaining employed and the vastness of the investment that such robotization entails. This limits the extent of robotization. However, the Covid pandemic has not left employers with any option other than to undertake a large investment in installation of robots at the workplace: the only way to avoid disruptions in the workplace is to ensure physical distancing of human labour by replacing human–human contact with human–robot contact. This results in a significant lowering of labour intensity in production.

Once the massive investment in robots has been made there is no going back to a higher labour intensity once the pandemic subsides: it would make sense to extract the almost free flow of services from the massive stock of robots in place rather than hire labour at a significant cost to provide these services. Thus, the pandemic would result in a permanent lowering of labour intensity.

There would be some positive effects on employment of the type I have already discussed at length – those resulting from increase in the scale of output of those sectors undergoing robotization in response to the pandemic and counteracting the decline in the amount of labour used to produce a unit of output, as well as others associated with marketing and sales of enhanced output.

It is also true that many of those displaced from their jobs as a result of Covid-induced robotization will eventually find new ones. But a significant interval of time would in most cases separate displacement from jobs and employment in new ones: the formation of mutually beneficial employer–employee relationships has to be preceded by time-consuming search on the part of employers and employees and consolidated through a process of training. The mentioned interval of time would thus see significantly lower levels of employment and a resulting recession in aggregate demand unless measures are taken by the government to revive demand.

As I am in a race with deadlines trying to finish this book, U.S. President Joseph Biden has announced *The American Jobs Plan* to counter the unemployment arising from Covid 19, which has to be a result of Covid-induced robotization and establishments downsizing and folding up because of receding demand. It is inevitably the large firms that manage to robotize and stay afloat and the small ones, with their poor capacity to raise credit and make burdensome interest payments, which respond to lowered demand and the difficulties of continuing production in the traditional way by folding up or downsizing to a skeletal state.

In her contribution to *USA Today* in the middle of April, Kamala Harris (2021), the U.S. Vice President, draws our attention to the severe problem of unemployment in the United States as it emerges from the pandemic: ten million, many of them blue-collar workers, unemployed. President Biden's *American Jobs Plan*, she said, would not only upgrade national infrastructure but in doing so provide good jobs to the unemployed. While existing skill sets would be utilized, necessary skills training – which this book emphasizes is needed so that workers can get new jobs after getting displaced from existing ones due to robotization – would be provided to make employees more suitable for available jobs.

To sum up, the *mantra* needed for economic recovery from the shock dealt by the Covid pandemic is to shore up consumption expenditure – in order to check economic inequality, ward off impoverishment, and counteract recession in demand – through a minimum income plan and make people employable and economically self-sufficient through skills training after being rendered unemployed.

What do research studies tell us about the economic impact of robotization?

It is generally difficult to find a study which highlights all the possible major consequences of robotization. But when we combine insights from various studies, a useful mosaic emerges which is very useful for policy. Some of these studies are empirical and others are theoretical. There is no attempt in this chapter to discuss these studies, with proper attribution, in detail as this has been done in the forthcoming chapters. Instead, what is done is to juxtapose the findings of various studies so as to generate a comprehensive picture of the overall impact of robotization.

Robots, capable of functioning without human intervention for long periods of time, have been introduced to take over many tasks which were earlier undertaken by humans. Thus, they have displaced humans from jobs and reduced wages. According to recent studies on robotization which have combined theoretical frameworks with empirical exercises the resulting unemployment has been alleviated, but not neutralized, by non-robotized industries hiring more labour at the mentioned reduced wage or production units trying to benefit from the consequent increase in labour productivity by using more humans for carrying out non-robotized tasks.

Economists have pointed out that while automation results in displacement of humans from jobs it also results in opportunities for new jobs for coordinating the use of machines to perform new tasks and thereby generate new products and services. The mentioned opportunity might be short-lived if a spurt of automation is quickly followed by another so that the mentioned generation of new products and services can be accomplished almost purely by machines. This would of course be determined by the flexibility of technology for generating inventions and innovations on the basis of machines. In recent times, the Robot Operating System is one such flexible technology which has come into use; similarly, Microsoft Windows is an operating system which can host various kinds of software, targeted to achieve varied objectives, on desktop computers and laptops.

Thus, we can say that the window for generating new job titles, which can coordinate recently invented machines, has narrowed. Given that this window represents a force which counteracts the negative impact of automation on human employment, the "narrowing" represents a diminution of employment opportunities for adult humans over time. A cataloguing of experiences of various countries show that unemployment and wage declines characterize both skilled and unskilled labour. At the same time, there do exist occupations in both skilled and non-skilled segments of the occupational structure which should remain the preserve of humans over a long period of time: for example, those that require soft and fuzzy skills such as the ability to deal with unforeseen contingencies or express appropriate emotions in dealing with human customers (managers, those providing geriatric care, those serving in the hospitality sector and entertainment industry, etc.) and jobs requiring complex limb movements such as those held by janitors and cleaning ladies.

While the decline in employment caused by robotization adversely impacts aggregate demand and therefore the rate of growth of per capita income, there are other ways in which AI tends to stimulate the growth of per capita income: the speed of conducting research and dissemination of results from research, an input into further research, increases. This happens because of the internet cutting the search time for accessing information/data/research, and the constant improvement of computers over time which enhances the speed of processing information/data as well as the quantity of data which can be processed at a point of time. An increase in the speed of conducting research implies that technological progress becomes faster. Lastly, the internet becoming a part and parcel of our lives implies that the transaction costs associated with various activities have shrunk and this has contributed to people undertaking transactions which they would have never engaged in before. These range from purchases, through the internet, of goods and services which were earlier not undertaken as the required expenditure of money and time on travel was forbidding, to transactions that represent an increase in the consumption of government services, such as provision of passports for travel, nowadays associated with a transaction cost of a few minutes spent on the internet as compared to many hours earlier. The increase in the number of transactions brought about by the AI revolution enhances welfare per capita; if the right monetary value is imputed to this enhancement of transactions, it would be reflected in a significant increase in the per capita national product of the country.

It is, however, seen that growth rates of per capita income, as computed, are lower in the AI era than in the pre-AI era. Some part of this reduction in the growth rate of per capita income is an outcome of contraction in certain sectors of the economy: meetings going virtual tend to depress the demand for fuel and services provided by the travel and hospitality industry; digitization of data capturing sound, images, and text implies that the material basis of the movie, music, and publishing industry has shrunk, which is bad news for the sectors of the economy that provide these materials. On the other hand, there are many developments in the AI age which have emerged as significant contributors to economic welfare but have not resulted in comparable increases in estimates of the gross domestic product (GDP): for example, free services such as those provided by email and social media accounts and websites which provide access to video and music, either free or for a nominal charge; free access to leading magazines, newspapers, journals and databases, blogs, etc. provided through the internet. Thus, we can safely conclude that a comparison of the growth rates of per capita income in the AI and pre-AI era is not appropriate, given that GDP is computed on the basis of revenue streams and such computation bypasses the ever-increasing volume and value of free services in the AI age.

At the same time, it must be recognized that the AI revolution does tend to reduce employment and through it aggregate demand. This in turn generates tendencies which can adversely affect per capita GDP and therefore its growth. Government policy thus needs to concentrate on how new employment can be generated through the identification of sectors characterized by unmet demand as well as the reskilling of workers rendered unemployed to enable them to perform jobs for which there is demand. At the same time, there are channels through which AI has a salutary impact on aggregate demand: by providing services to workers which make their workload lighter and thereby catalyse the shortening of the work week, it can enhance the time available for consumption, and by shortening the amount of time needed to complete transactions, it enhances the number of transactions that can be completed in a given swathe of time and reduces transaction costs, which also makes it worthwhile for consumers/ beneficiaries to undertake an enhanced number of transactions.

Using policy to regulate the adoption of AI while reaping its benefits

The use of robots and AI has not only greatly enhanced the efficiency with which several tasks in production processes are performed but will continue to do so. As AI continues to develop as a science and its application becomes broader, it also enables research to be faster and make contributions to the stock of knowledge, inventions, and innovations, which would not have been possible without this development. Thus, development in AI has had and will continue to have a significant positive impact on technological progress: old tasks will be accomplished at unprecedented speed; accomplishment of new tasks and associated production of new goods and services is being made possible; and lifesaving technologies using AI as a basis are being devised to counter deadly pathogens and adequately feed the expanding population of a teeming planet.

But at the same time, the wider use of AI does pose challenges to policy makers as AI is in many cases a substitute for human labour. Thus, it can destroy jobs and bring about a reduction in wages. For AI to be adopted in an economic sector, it has to be able to contribute to the accomplishment of one or more tasks constituting the production process and do so in a manner which is more profitable for the entrepreneur than the use of human labour. There, however, can be major problems arising out of the rational instincts of entrepreneurs: when use of AI to replace human labour becomes profitable in a number of sectors at the same time, major bulges in unemployment can occur. Unemployment is unjust as it robs those affected suddenly of the means to earn a livelihood. It might also result in a deepening of poverty. But it also affects demand facing various sectors of the economy, plunging them into recession and forcing them to lay off workers, which deepens initial unemployment.

It is important to note that the government can ensure that the mentioned large bulges do not take place: the idea is to keep an eye out for economic sectors in which labour displacing use of AI has recently become profitable and therefore its incorporation into the production process is imminent. This allows the government to put in place an array of robot taxes at a point in time such that it is economically feasible for only a few sectors to adopt robotization at that given time. If the number of sectors which have turned suitable for robotization is large, then the mentioned array should generate a queue: those facing large taxes will have to wait for a while to undertake robotization till the government reduces these taxes. The idea of course is to eventually allow all profit generating entry of AI into production but not too much entry at any given point in time. When bulges in unemployment are avoided in this manner, unemployment rates are kept within limits and impoverishment lowered, but also the likelihood of major recessions of demand in various economic sectors which lead to further spiralling of unemployment is greatly reduced.

The robot tax would also generate significant amounts of revenue from those sectors which find it profitable to use robots in production in spite of such use being made relatively expensive in comparison to the employment of human labour through this instrument of taxation. This revenue can be used to supplement the impact of the "bulge prevention" discussed in the last paragraph: basic or minimum incomes can be financed to not only shore up the fortunes of those rendered unemployed by AI but also check the mentioned recession of demand and therefore the spiralling of unemployment; appropriate reskilling of those rendered unemployed can enable them to find jobs for which there is demand; and ownership of physical and financial capital by such unemployed workers can be facilitated to counteract the uncertainty characterizing returns from human capital in the AI age.

The implementation of the robot taxes, which includes fixing their sector-specific rates, would have to be done on a global basis: if the rates for any given sector are different across countries, capital flight would result, and anticipating this behaviour, countries would indulge in competition taking the sectoral rate in all countries to zero. Thus, what is needed is a global agreement specifying an array of sector-specific rates, one rate for each sector, for implementation all over the globe; the revenues from these taxes can be used to build a global fund to provide basic or minimum incomes, undertake the mentioned reskilling, and endow displaced labour with ownership of physical and financial capital. Basic or minimum income levels facilitated would inevitably have to vary across countries, keeping in mind the huge variation in living standards/per capita incomes/wage rates across countries.

If the global fund actually takes shape, the allocation of this fund for various mentioned purposes and the relative agreed upon magnitudes of basic income across countries would be the outcome of negotiations and diplomacy. But the immense potential for this fund to alleviate current pain and avoid future misery is enough reason for the world's leaders to think about it. AI after all is both a global public good and a global public bad: as AI develops through research and learning-by-doing, the advances become available for improving and sometimes even revolutionizing production processes all over the world; at the same time, the incorporation of AI into production processes in one corner of the world can destroy the global markets for goods and services produced by traditional means in other locations; these various acts of destruction can add up and generate economic fires of mutually reinforcing unemployment, impoverishment, and recession. If the characteristics of climate change have earned it the label of a global public bad and resulted in various countries of the world contributing towards the building of a *Green Climate Fund*, there is no reason why a global fund for combating the adverse consequences of robotization, while benefitting from its gradual global adoption, should not be built.

Note

1 Useful discussions on this topic are present inMcBratney et al. (2005) and McWhelan et al. (2003).

2

AN INTRODUCTION TO ROBOTS AND THEIR ECONOMIC SIGNIFICANCE

The mission of this book

Historical background

Around 1779, British artisans, mainly weavers and textile workers, began to feel outcompeted by early textile factories in which relatively unskilled labour operated machinery such as mechanized looms and knitting frames. These machines were pushing the skills of these artisans, assimilated through years of apprenticeship, towards redundancy. For example, machines allowed workers to produce knitted goods about 100 times faster than by hand. Desperation gave rise to weavers breaking into factories and smashing textile machinery, following the example of a young artisan, Ned Ludd, rumoured to be one of the first to have wrecked a textile apparatus. These rampaging weavers called themselves Luddites (see Christopher Klein (History website) for details).

The Luddites were probably inspired by the success of what historian Eric Hobsbawm (1952) referred to as "collective bargaining by riot": textile workers and coal miners in the period between 1715 and 1765 disrupting the process of production to secure a raise in wages. They would have not objected to the use of new machinery had they been appointed to operate these with a raise in their wages; but the revolution had made their skills unnecessary for production and the merchants and entrepreneurs preferred to run their factories with unskilled labour hired at low wages.

The movement took time to pick up steam, with the first major instances of machine breaking taking place in 1811 in Nottingham about 30 years after its initiation, but then it spread like wildfire across the English countryside. The government responded by dispatching a sizeable contingent of troops and incarcerating prominent Luddites after decreeing machine breaking a punishable offence. By the end of 1813 the movement had been effectively suppressed.

DOI: 10.4324/9780429340611-2

The fears about machinery, not unexpectedly, remain to this day. Those exchanging their labour services for the means to a livelihood obviously feel threatened by machinery that can replace them or render their role in the production process insignificant. At the same time, it is obvious that capitalists welcome the introduction of productivity-enhancing machinery with a view to enhancing profits while trimming the workforce, given the problems of managing human labour. The introduction of robots, which have been gaining in popularity in the 21st century, marks a watershed moment in the history of capitalism. Defined as machines which can mostly operate themselves without human interference over long periods of time, complementarity between man and machinery is expected to weaken significantly and give rise to greater substitutability.

Consider a national survey conducted by Pew Research Center in the United States between June 10 and July 12, 2015, among 2,001 adults (see Smith, 2016, for details). When asked about the threat posed by computers and robots, one-in-ten workers were concerned about losing their current jobs due to workforce automation. According to them, competition from lower-paid human workers, often a consequence of new machines making human skills redundant, posed a more immediate worry. As much as two-thirds of Americans however anticipated that in 50 years robots and computers will be doing much of the work currently done by humans.

Two other polls of Americans in the workplace exist. A CNBC/SurveyMonkey poll conducted four years after the Pew Research Center Survey revealed that as much as 27% of the workers felt that technology was threatening their jobs, much higher than the earlier figure of 10%. Another poll, the Fortune/SurveyMonkey poll conducted in 2018 found that 72% of people expected artificial intelligence to take away more jobs than it created in the next ten years.

But the fears of those engaged in the daily business of earning a living and looking after their families are not the only fears regarding robots. What is more worrying are the doomsday predictions made by experts on the basis of a study of labour markets and how these are being impacted by the emergence of robots. Martin Ford (2015) in his book chillingly entitled *Rise of the Robots: Technology and the Threat of a Jobless Future* warns of an age, not very far from the present, characterized by few human jobs and immense inequality. Futurist Jerry Kaplan estimates that 90% of Americans will lose their jobs to robots (see Bowles, 2016) though he seems to be optimistic that displaced humans will pick up new jobs.

This book is motivated by the fears we have about robots taking away our jobs, our incomes, and happiness. Apprehensions about future unemployment and income inequality are however judged for their validity through an economist's lens. In this section, I discuss whether the human fear of machines has been justified by historical experiences and hard data regarding these experiences, and then initiate a discussion on the human fear of robots, a specific type of machine. The latter leads up to certain questions posed at the end of the section, with the rest of the book devoted to the voyage that helps the readers and author discover answers.

The human fear of machines: mostly unfounded?

Human fears about automation arise from the possibility that humans would soon lose the significant roles in the production process they currently enjoy to machines. Thoughts of this possibility arouse anxiety as for most human beings, consumption is determined by earnings from labour[1].

In other words, the argument goes as follows: man works in order to consume; if the demand for human work is replaced by the demand for machine labour, then for much of the population, unemployment and poor quality of human life would follow. Such fears have been expressed ever since the 18th century when the industrial revolution ushered in mass mechanized production. Yet the prophets of doom turned out to be wrong, though it must be admitted that the economic expansion which accompanied the industrial revolution was marked by a large number of jobs being destroyed while many new ones were created. First, most production required collaboration between humans and machines. Second, even though the amount of output that an hour of human labour, aided by tools, could produce increased enormously as a result of the revolution, the expected negative impact on employment did not materialize as many items earlier consumed only by the elite gradually became objects of consumption by the masses or at least a large chunk of the population, and product variety expanded with the help of innovation and invention.[2]

That the industrial revolution was fed by migrant labour from the countryside, liberated by the mechanization of agricultural production, is well known. It played a key role not only in enhancing consumption of the masses and making it more varied but in preventing unemployment caused by technological progress in agriculture. There was of course a symbiosis between employment creation in industry and the broadening of the consumption base which resulted in the lower and middle classes not only consuming an increasing number of goods but also increasing amounts of some of the goods they had already consumed: employment in jobs associated with more advanced technologies and mechanization providing the higher incomes needed for such consumption.

Automation of course continued into the 20th century, but yet again the fears of mass unemployment of a long-term nature proved to be unfounded. Consider all durable goods which started off as items of consumption of the elite in the 20th century and over time came to be consumed by vast chunks of population in developed and even developing countries: cars (see Box 2.1), televisions (see Box 2.2), washing machines, dish washers, refrigerators, microwave ovens, air conditioners, mobile phones, various types of personal computers, etc.

As consumption of these goods increased, there was a burgeoning of factories producing these goods, thus facilitating increasing levels of employment of human labour. This helped to compensate for the loss in employment caused by automation in agriculture, manufacturing, and services, producing conventional items such as food grains, cloth and garments, traditional transport services, etc. The motorcar and its cousins in the public transport sector, the train, tram car, and motorbus, made movement of goods and people less intensive in human labour but opened up new vistas in the economy so that the overall implications for human employment were positive. The television carried the moving picture and embodied entertainment to the doorstep of every household and ushered in a huge expansion of the news and entertainment industry. At the same time, the attractions of the "large screen" and its life-size images drew people to newly constructed movie halls all over the globe. Labour employed by the news and entertainment industry grew manyfold. It must be remembered that television sets are consumer durables which are owned at the level of the household and therefore their manufacture alone constitutes a major source of employment. With approximately 1.5 billion households[3] in the world today, the major sources of global demand for television sets are (a) replacement of television sets which have outlived their use due to obsolescence in the face of a rapidly improving technology or depreciation; and (b) income growth enabling many developing country households in the mentioned population to become owners of television sets for the first time. A similar argument can be made regarding demand for other consumer durables. Population growth, the tendency for household sizes to become smaller with economic growth, and income growth which enables households to own valuable consumer durables will thus continue to be the sources of growth in such demand.

The story of the television is that of a consumer durable invented in the 20th century which enabled the beaming of services such as news and entertainment directly into homes. Households gradually came to consider these services and their deliverers, the television sets, as essential. A single durable thus came to contribute significantly to the economy by facilitating employment in service production and delivery, as well as manufacture, apart from being a major source of value through entertainment, human capital formation, and vital information (see Box 2.2). There seems to be no sign of the contribution of this 20th century invention to the world economy waning. But the story of the television also serves as a parable offering an important lesson: predictions of massive unemployment brought about by the current revolution in artificial intelligence should be made with caution as there is surely scope for new durables such as robot butlers to provide a stimulus to the economy and generate employment comparable to that facilitated by the television.

Are our experiences with conventional machines poor predictors of future experiences with robots?

Experts however argue that robotization is different from previous automation as robots can function without human interference for a long time, for example, a garment-producing robot can transform cloth into garment in a matter of minutes without a human worker even moving a finger. The fear is that if robotization outpaces increase in product variety, there could be net adverse implications for human employment of an enduring nature.

BOX 2.1: THE MOTORCAR: MORE BOON THAN BANE

In 1870, the functioning of the American economy was extremely dependent on horses. Every family relied on the horse directly or indirectly: as a means of transport, as a deliverer of important consumer goods, and as a consumer of hay produced by it if it was a farm household. The American human population of 43 million derived its livelihood on the basis of support from an equine population of 8.6 million, one horse for every five persons. Interestingly, the industrial revolution was gathering momentum at this time. Powering this enormous momentum was the enhanced use of horses as a means of transportation. Horses were being used to pull passenger carriages and delivery carts. Other uses of horses lay in imparting mobility to brewery wagons, city vehicles, omnibuses, etc.

The increase in economic activity came at a cost: the horses were a significant source of pollution. For example, in 1908, New York's 120,000 horses swamped the city streets with 60,000 gallons of smelly urine and 2.5 million pounds of excreta every day. The adverse implications for health and sanitation were enormous. Right at this time the technology for making the automobile or the modern-day motorcar became available. This kind of transport had obvious advantages: no liquid and solid waste and a much higher speed. Motivated by the potential for improvement, the transition from horses to automobiles was completed in a very short period of time. As horses vanished, they took away a large number of jobs associated with their use. For example, in the period between 1890 and 1920, the number of companies building horse-drawn carriages in the United States went down from 13,800 to 90. However, the mentioned job destruction was almost immediately neutralized by the job creation that accompanied the rise of the automobile industry.

Invented in Germany and France towards the end of the 19th century, the motorcar became quite popular in the United States in quick time, probably because of the need for travel between scattered and isolated settlements and a high incidence of prosperity in its population. In 1913, the United States produced 485,000 of the world-total of 606,124 motor vehicles. The U.S. and world populations of 97 million (Demographia: website) and 1.8 billion (Worldometer: website), respectively, meant that there was a car for approximately every 200 people in the United States and every 3,000 people in the world. If one left out the U.S. population from the world population, then there was a car for every 12,000 people.

In the subsequent years, with improvements in technology shoring up the supply side through quality improvements and cost reductions and economic growth enhancing demand, there was a huge boom in automobile ownership: in 1980, 87.2% of American households owned at least one motor vehicle each while 51.5% owned more than one. From a car for every 3,000 people in 1913, car ownership in the world had become enormously more common at a car for less than four persons.

In spite of mounting competition from Japanese car manufacturers, the U.S. manufacture peaked at 12.87 million units in 1978. In 1982 the automobile industry provided one out of every six jobs in the United States. It was the principal source of demand for the petroleum industry and a very major consumer of steel and many other industrial products.

The automobile spurred the demand for tourism and related industries such as service stations and roadside restaurants and motels. It also connected population in far flung areas to important services such as quality medical care and schooling. With demand rising for various products and services, it became economical to expand the network of streets and highways. The resulting improvement in infrastructure improved quality of life and was a boon to industry.

The story of the automobile industry is an illustration of how automation can lead to varied job creation in other industries and stimulate economic growth; the low labour intensity of a rising industry is thus a poor barometer of its potential for creating employment.

Major references used: History (April 26, 2010),
Smith and Browne (December 21, 2017)

Others argue that robotization will adversely affect employment only in those occupations where the tasks at hand can be broken down into clearly delineated steps and there is no need for the exercise of discretion during the process of production. As we shall see later in the chapter on machine learning, it is now possible for computers to learn tasks which humans perform without being able to delineate the steps. These include tasks such as face recognition or recognition of fruits, vegetables, or animals.

Other experts correctly argue that as perfect certainty about the working environment clearly does not characterize many occupations, humans will continue to be gainfully employed in these as they are much better at dealing with uncertainty than robots. Humans also are capable of employing common sense, emotional intelligence, and varied background knowledge accumulated at different points in their lives to complete tasks. Examples of such workers are caregivers and maids, paid companions, and managers who all use a lot of soft or fuzzy skills for which, as of now, computer programmes cannot be written. For example, a cleaning lady can use experiences and challenges faced in regard to cleaning rooms as well as dealing with customers to improve the quality of her services. The width of such knowledge is not available to a robot. The ability of people in these occupations to "empathize" is a characteristic which is much valued by consumers; again, it is very difficult for a robot who has not had the human experience to effectively replace humans as offerers of empathy.

BOX 2.2: THE TELEVISION: A WINDOW TO THE ENTERTAINMENT ECONOMY

In the United States, 96.1% of households owned television sets in 2019, a penetration rate which is way higher than that observed in the 1950s and 1960s. Using information provided by Wiegand (2016) and data on household size (Statista, 2020), we can infer that at the start of the 1950s, less than 1 million households owned the recently invented television sets but by the end of the decade television ownership characterized 18 million households or 35% of the total number of households in the United States. Clearly there was a very high growth of ownership in the 1950s and 1960s.

While the household use of television has reached almost 100% in the developed countries, in developing countries it is still growing. As televisions are consumer durables, the world market for the "small screen" will continue to expand as developing countries in Asia, Africa, and Latin America climb up the income ladder. Consider television ownership in India, a middle-income country with about a fifth of its population still living below the international poverty line. In 1993–1994, 40% of all urban households, accounting for a quarter of its population, and 7% of rural households possessed television sets (National Sample Survey Organisation, 1997). A back of the envelope calculation yields that about 15% of all Indian households owned television sets. By 2009–2010, 42% of rural households and 76% of urban households, now accounting for approximately a third of its population, possessed television sets (National Sample Survey Organisation, 2012). Calculation reveals that slightly more than 50% of Indian households now owned television sets, thus associated with an increase of 35 percentage points in a mere 16 years. The penetration rate is bound to increase over time going by the development experience of present day developed countries.

It is not difficult to imagine what the economic benefits flowing from the television industry, apart from those related to employment and input demand associated with manufacture, are. These are the consumer surpluses associated with the viewing of entertainment, the human capital formation resulting from educational programmes aired on television, and the reduction of information deficits in regard to politics and current affairs as well as consumer and business opportunities. In economics, consumer surplus from a product is defined as the excess of the amount people would be willing to pay for a product over what they actually pay for it.

The consumer surpluses from television programmes are obviously massive given the large audience and free viewing: imagine the cost that a person willingly incurs to be a spectator at a sports event (a cricket or football match, the opening ceremony of the Olympics, etc.), an expenditure that is obviously not greater than the monetized benefit from viewing the sports event, given that he is rational; now imagine the surpluses (net benefits) that emerge from millions, often hundreds of millions, watching this event on television with no cost incurred by the broadcaster as more spectators join the broadcast. The magnitude of these surpluses is indicated to an extent by the advertisement revenues generated for the broadcaster. For example, NBC Universal came out with a statement in the end of 2019 that it had received more than $1 billion in national advertising commitments for the 2020 Tokyo Olympics (Graham, 2019), which of course never took place as scheduled. The cost of advertisements ranged between 1 million and 100 million. Nowadays, with digital platforms also emerging, companies can buy advertising time on television programmes taking place in real time as well as the runs and reruns of these programmes on digital platforms.

The opportunities from human capital formation occurring through television programmes, and in recent times through programmes on digital platforms, are massive. Economists call this a case of zero marginal cost – when one more viewer is added to the audience, the broadcaster incurs no increase in cost. In fact, it is in the interest of the broadcaster to enhance the size of the audience to the extent possible as more eyeballs translate into more revenue from advertisements. Unlike education in classrooms, there are no capacity constraints nor any need for learners to be in the physical proximity of the instructor. The advantages of these programmes are therefore as follows: significant human capital formation in a very large number of viewers; large advertisement revenues for broadcasters; an opportunity for advertisers to showcase their products to a larger audience yielding a value exceeding the amount paid to broadcasters.

Finally, television programmes and advertisements surely ease the information constraints facing potential investors and consumers. The resulting increments in investor and consumer activity are very significant, given the large viewership.

To appreciate a shortcoming of human–robot interactions, consider this interaction – described in italics – between a human owner and his very contemporary digital assistant, Robota. The digital assistant bears a fictitious name as my intention is to highlight the shortcomings that are common to all contemporary digital assistants or house robots without in anyway damaging the reputation of a specific brand.

On a calm afternoon when I ask my digital assistant Robota to play some Beethoven tunes, she obliges to my great joy without my having to lift my tired body out of bed. But when I ask her about whether she believes in God, she is not able to profess any opinion. Robota thus turns out be an assistant, not a companion.

Note that the above conversation is the dramatized equivalent of a recent interaction with a digital assistant. It captures the ability of robots to perform mechanical jobs efficiently and yet demonstrates their inability to engage in conversation voicing opinions, and therefore indulge the human thirst for debate and discussion on abstract and normative issues. The mentioned ability along with emotion and empathy are qualities that a human values highly in his fellow beings, as these form much of the bedrock of human relationships and social interactions. As long as robots do not possess such attributes or are far poorer in regard to these attributes than humans, human–robot interactions will not be able to totally or even significantly replace human relationships or social interactions. Since robots do not undergo the human experience, the number of meaningful issues on which they can have conversations with humans is limited. This includes abstract issues. Thus, I expect that many occupations, such as those in the hospitality and caregiving industry, will continue to be manned by humans.

However, it must be conceded that interactions with contemporary robots cannot tell us what the distant future holds for us. Experiments aimed at replication of neural networks and other artificial equivalents of the human brain are reportedly in full swing. Details of these experiments are obviously guarded very carefully and one can only comment on the products that are on display in the public domain. In the distant future, humanoids or robots which not only look like humans but match them in regard to the range and intensity of emotions, mental and physical flexibility, and abstract thought might well become a part of our daily lives. However, we can be very sure that firms which are on the brink of such inventions, if there are any, will not be able to market these products for common use in the near future because of the lengthy process of marketing and licensing associated with the market launch of a product having such significant implications. This can be concluded because of the experience with driverless cars which have existed for a long time now but are not yet a part of the daily lives of humans in most parts of the globe.

Human workers have always felt threatened by artificial intelligence. When computers came into use in a big way in the last two decades of the 20th century it was felt that they would take our jobs. In fact, in the late 1970s and 1980s when unemployment spurted rapidly and computer technology was embraced by every sector of industry in most developed countries, many critics linked the two. *The Age* newspaper in Australia presented an article by Steve Harris in 1983, which listed a large number of occupations that would see unemployment: proofreaders, library assistants, mail couriers, accountants, financial analysts, administrators, etc.[4] The computer (silicon chip) was dubbed the contraceptive pill for the labour force.

But jobs created by computers were not restricted to the manufacture, packaging, retailing, and servicing of the labour replacing machines, as the critics suggested. As computers came in, employees such as accountants and secretaries were put to tasks that were earlier considered to be too time consuming. Australian data shows that there were in fact minor shortages for these two occupations, negating the myth that these occupations were characterized by an increase in unemployment due to computerization. With bookkeeping reduced, the accountant could process data to identify sales and customer trends and generate reports on profit levels and cost structures. Secretaries, too, could spend more time in filtering information for their bosses, personnel management, and communication.

There are other reasons why computerization and related automation might result in employment generation offsetting the job destruction associated with the lowering of labour intensity in a large number of productive activities. First, the embrace of robotization means that production relies more on machines than on men, with innovation over time leading to quality improvements and cost reductions in these machines. This possibility of (a) greater mechanization as well as (b) machines such as robots and computers becoming better with time, with the first trend often a consequence of the second, goes hand in hand with greater price competition. The resulting lower prices lead to greater demand for robots and computers as well as the products these help to produce and consequently greater volumes of production and magnitudes of employment. This happened, for example, in the watch industry. Price cuts also mean that more income can be spent on other products – some of which, such as those in the hospitality industry, being expected to remain quite labour intensive in the foreseeable future – resulting in greater employment in these product lines.

These analytical arguments are backed by empirical evidence. For example, in the period 1975–1983, it is not the countries which experienced the greatest computerization that saw the greatest increase in unemployment. One of the reasons was that many new jobs were created through the emergence of new sectors such as information technology. The scope for humans to compute, communicate, and transact increased immensely and the market was consequently flooded with new products such as ready to use software and new data-driven services. Employment of humans was needed not only to produce these new products but also to market them. Besides, computers themselves had to be repaired, creating a lot of new jobs.

Articulating the mission

While the lessons from history are important, our past experiences with computers and robots, especially those captured by trends in employment over time, might not on their own be the ideal predictors of what is going to happen in the future, even in the next couple of decades: research in artificial intelligence implies that robots are evolving at a rapid pace with the frontiers of production being pushed back rapidly. Any broad predictions in regard to economic variables thus need to be based on both the mentioned trends as well as recent developments in the field of robotic innovation. This book does exactly that in answering the important question it poses: what tendencies will be generated in the foreseeable future by the march of artificial intelligence (AI) technology in regard to the creation and destruction of jobs and change in welfare levels, and what is the role of policy in checking undesirable tendencies and enhancing the impact of desirable tendencies so that the quality of life of the common man improves, poverty alleviation occurs at a swift pace, significant economic growth occurs, and income inequalities are bridged rather than enhanced.

The book does not generate numerical estimates though it uses the following as inputs: empirical studies, often prediction-based, along with theoretical models, descriptive studies of life around us, and trends in technology. The result is an open-ended analysis entertaining diverse possibilities, with the focus on how policy can be used to generate improved economic outcomes in the AI age. As the ensuing pages will reveal, there is much that the governments of the world can do to reduce human pain in the course of the transition from manually operated machines to robots. What even the immediate future holds for us would critically depend on the measures undertaken by national governments unilaterally or more importantly through multilateral agreements. The book seeks to therefore contribute to policies that would be gamechangers in the robotized world.

An introduction to history of robotization and concepts to be used in the book

This introductory chapter thus sets the stage for what is to follow in the book. Having already delineated the central objective of the book, this chapter goes on to provide a brief history of robotization, which for a major part is a history of humans imagining and fantasizing about robots. This will be followed by a discussion of necessary concepts and definitions of specific terms such as "artificial intelligence" and "robots", given that scientific definitions often differ from the common man's understanding of these terms generated from movies and novels. Finally, this chapter shall provide the outline for the rest of the book and show how the book has been structured to build a systematic body of knowledge that facilitates the attainment of its central objective.

History of robotization

The Greeks were probably the first to think of robots. One of the products of their imagination was *Talos*, a bronze man whose blood was made out of lead, considered to be a divine fluid (Encyclopedia Mythica). The next significant landmark in thinking about robots was due to Hero of Alexandria (A.D. 10–70) who conceived of various *automata* – mechanical devices capable of performing a range of functions according to a precoded set of instructions – including statues which could pour wine. Whether such mechanical devices actually existed or were purely imagined can be debated (see Porro, 1589, for details).

In the medieval times, life size human dolls capable of some movement through hidden mechanisms were used to induce peasant worshippers in the church to believe in a higher power. The Arab polymath Al-Jazari (1136–1206) left texts pointing to similar dolls: a band of musical robots and a robotic waitress that could serve drinks. Around 500 years back, Leonardo da Vinci came up with the design of a machine which resembled the human body's mechanism (Illumin, 2018). He is also credited with designing a humanoid knight clad in a German-Italian suit of armour capable of some movements. NASA roboticist Mark Rosheim published an independent study of this robot after studying Vinci's notes on the same and reconstructed it for a BBC documentary in 2002. Vinci also documented the design of a robotic lion. Around the same time, a clockwork monk capable of moving its lips and arms in silent prayer was invented in Europe, though there is a lot of uncertainty about who the inventor was. It is still in working condition today and stands in the Smithsonian, Washington D.C. where it has been studied by one of the museum's conservators to reveal the complex mechanisms within (Davis, 2012).

Robotization as a concept is closely linked with computerization, which is basically devoted to producing superhuman abilities in carrying out cognitive tasks that humans can perform, and other abilities, which humans do not possess such as creation and scanning of images. The year 1645, when Blaise Pascal invented the *Pascaline*,[5] marked the starting point for inventions that tried to match and then outmatch human capabilities of addition, subtraction, and multiplication. Only 21 years later, Samuel Morland invented a pocket version of the *Pascaline*. This marked the beginning of the time trail that would lead to the invention of modern computers, which keep on getting remodelled and upgraded with the passage of time.

Jacques de Vaucanson, a Frenchman born in 1709, invented two robots in successive years starting 1738: a life size flute player with a repertoire of twelve pieces and "The Duck" which could flap wings, eat, and digest grains.[6] Vaucanson was followed by Charles Babbage (1791–1871), a scientist of the Victorian era who developed the early foundations of computer science, which would endow robots with human-like cognitive skills, especially in regard to mechanical calculations, and others such as Ada Lovelace who explored the possibility of creating non-human skills such as developing images. Over the course of the 19th century machines capable of performing complex tasks, hitherto carried out by humans, came into being: one that could be programmed to create designs; a Steam Man,[7] an armoured gas boiler, which could walk independently at a rate of nine miles per hour;[8] and later an Electric Man for pulling wheeled carts.[9]

In 1921, the term "robot" entered the lexicon with the staging of a play, *Rossum's Universal Robots*, written by the Czech writer Karel Capek. In the play, man invents the robot to replace him.[10] The term "robot" is derived from the Czech word *robota*, meaning serf or labourer (Hockstein et al., 2007). The robots in Capek's story looked like humans but were made of chemical batter, whereas in reality they are made out of metal. However, they, as pictured by the playwright, were more violent than humans and ended up killing them.[11]

The play probably helped to trigger human imagination about machines which could perform a range of human functions relating to cognition and in certain cases fulfil the role of humanoids, artificially created machines, which had the physical as well as functional characteristics of humans. This process of humans imagining artificial versions of themselves was given full flight by science fiction writer and scientist, Isaac Asimov in 1941 who first used the word, "robotics" in his story *Liar*[12] to describe the technology of creating, studying, and using "robots". Since then, robots have become popular subjects in literary and cinematic fiction (Hockstein et al., 2007), sometimes as human companions (*Star Wars, The Jetsons, Bicentennial Man*), but often as man's enemy (*The Terminator, Stepford Wives, Blade Runner*, etc).

The developments in the real world in relation to robotics started running in parallel to man's imagination regarding robots in the second half of the 20th century, each probably reinforcing the other. In 1954, George Devol designed what is acknowledged as the first truly programmable robot and christened it UNIMATE for "Universal Automation".[13] The significance of his invention was recognized by Joseph Engelberger who, in 1961, formed the world's first robot company "Unimation Inc.", which stands for "universal automation", to commercialize Devol's invention. The Unimate 1900 series was the first of the robotic arms produced for factory automation.[14] It was used in auto welding. Even though it was fixed in space, it could make sense of its environment and was much faster than any human in the assigned task. The 1960s were also marked by the development of *Shakey*, which was mobile unlike Unimate but perceptive like it: a robot on wheels, which could navigate a complex environment.

Robotics took a huge leap forward in 1964 when artificial intelligence research laboratories were opened at Massachusetts Institute of Technology (M.I.T.); Stanford Research Institute (S.R.I.), Stanford University, and the University of Edinburgh. In the mid-1980s, Honda started a humanoid robotics programme. It developed *P3* which could walk quite steadily and shake hands and a more advanced humanoid, *Asimo*, which could even kick a ball.

I end this section here as a discussion on contemporary developments in robotics and their possible economic implications is conducted in Chapter 3. But before that discussion, it would be helpful to flesh out scientific definitions of basic concepts which are relevant for the book. These definitions are often at odds with the common man's understanding of these terms, catalyzed by movies and fiction.

Robotics: basic concepts

In this section, I look at basic concepts related to the subject of this book: artificial intelligence (AI), robots, and machine learning. At the outset, it is very important to understand what we mean by artificial intelligence. The term as translated refers to human-like intelligence generated in artificial objects. Much of what Herbert Simon (1995) said in this regard still remains relevant today. While Simon compartmentalized artificial intelligence into three categories, it is possible to collapse these into one definition of artificial intelligence: the ability of any machine or cluster of machines to replicate the outcomes of human intelligence. This can be done through computing machines (i) which function very differently from humans or (ii) those which can be designed through scientific efforts to mimic the functioning of the human brain (neural networks: see Box 2.3). With the progress of research in artificial intelligence led by human scientists, artificial intelligence should continue to broaden, as it indeed has over the past 50 years, to enable computers, robots, and avatars to perform a wide range of human tasks without human intervention. It is in this role that artificial intelligence poses a potential threat to the bread and butter of the common man.

BOX 2.3: NEURAL NETWORKS

Neural networks are computer programs which function in a manner similar to the network of neurons in our brain. These can solve problems and even learn from experiences encapsulated in data. The theoretical basis of neural networks was created through collaboration between a neurophysiologist, Warren McCulloch, and a mathematician, Walter Pitts, in 1943. Theory was applied to run a neural network for the first time by Belmont Farley and Wesley Clark in 1954.

Built like the brain, neural networks also try to imitate the brain's pattern recognition skills. Neural networks consist of layers of artificial neurons: an input layer which receives information/data from the environment, intermediate layers where processing of the information occurs, and the output layer which provides the solution to the problem posed.

Commercial applications of neural networks include investment decisions, recognition of handwriting and human faces, and even detection of bombs. Neural networks are trained using supervised learning, involving datasets which contain the input data as well as the results their processing should provide. This is uncannily similar to how children are nurtured by their parents, correcting them in their usage of words and recognition of concrete and abstract concepts.

Adapted from Vladimir Zwass (Encyclopaedia Britannica)

AI is exhibited by machines which possess one or more human traits – knowledge, reasoning, problem solving, perception, learning, planning, ability to manipulate and move objects[15] – though they might also possess non-human traits. A drone, for example, can move from place to place through aerial flight, which is not a human faculty, but is aware of the environment around it (perception), which is a human faculty. Thus, a drone possesses AI. AI therefore can endow machines with the power to replicate both human faculties, such as calculation and cognition, as well as non-human ones, say satellite imaging (Charniak and McDermott, 2009; Simon, 1995) though the initial definition of "artificial intelligence" was human intelligence produced in inanimate objects. AI ranges from "very narrow" to "fairly broad" and "very broad", exemplified, respectively, by a calculator, a computer, which can run several software programmes simultaneously, and a network of computers, which can replace a team of consultants devoted to project implementation by internalizing the functioning of the team. Over time, there has been a tendency for AI to become broader.

Robots can be considered as receptacles which store AI and possess the human faculty of being aware of what is going on in their surroundings, an ability which helps them to function autonomously. These, according to the *International Federation of Robotics* (I.F.R.), are machines that are "automatically controlled, reprogrammable, and multipurpose" (Acemoglu and Restrepo, 2017). While there is some ambiguity about the general definition of robots and some disagreements among roboticists, they agree on some general guidelines: a robot is intelligent, has a physical body, is capable of performing tasks autonomously, and is aware of its environment and capable of manipulating it.[16]

Note that autonomy implies that a robot is aware of the environment in which it functions. If the environment is marked by certainty, then a robot does not need to possess machine vision, the mechanical equivalent of "human vision". For example, movement of robots in an industry assembly line follows a strict time table and is so perfectly timed that robots do not even need "machine vision". Even without this faculty, robots can work harmoniously to manufacture or transport objects. The human contribution lies in programming such industrial robots and providing them an environment not marked by the uncertainty that humans themselves encounter. Such an environment is produced because the environment faced by a robot itself is a function of the precisely timed movements of other robots. Thus, consider n robots whose movements in a given production process follow a completely predictable time sequence with the movement of the *ith* robot taking place at time t_i and lasting exactly Δt_i. In such an environment, useful knowledge possessed by a robot of the environment around it only amounts to just knowing when it has to perform its task and for how long: awareness of the movements of other robots adds no value as it does not generate any additional positive input for work to be done in an environment marked by perfect certainty.

However, perception becomes important in an environment marked by unpredictable acts of nature, including that of human actors. In such an environment, an object which cannot perceive the unpredictable acts of nature is not capable of autonomous functioning. It is therefore not a robot. Thus, a plane which needs to be piloted around is not a robot. On the other hand, a drone, which can take off and land on its own with a sense of the objects around it, is a robot. This discussion therefore leads to the conclusion that whether an object is a robot depends on not only its own properties but that of the environment in which it operates.

Unlike humans, robots are capable of working in 24-hour shifts and can be reprogrammed easily to implement changes in manufacturing processes. Contrary to popular belief, robot programming need not always be done by humans: by just watching humans, software residing in a robot can often note the sequenced steps needed to perform jobs and then replicate the sequence. This is a type of what is known as "machine learning". But most importantly it seems that the cost of training robots is a fraction of the cost of training humans. For example, consider the Baxter robot manufactured by Boston-based Rethink Robotics. A manufacturer can train a single Baxter robot by conditioning its limb movements to do a variety of jobs. This knowledge is then propagated to other robots, numbering hundreds, using a USB device (Ford, 2015).[17]

Machine learning gets better as more and more data on humans handling jobs emerges, thus providing clear directions for handling a limited number of contingencies. However, machine-learning algorithms for replacing humans are clearly difficult to construct if the humans being considered are those whose jobs involve dealing with a large number of contingencies, especially ones which are difficult to anticipate in advance and therefore require soft or fuzzy skills. This is a major reason for recently developed humanoids not being able to emote, collaborate, and interact socially as well as humans (Stephane et al., 2016).

The layout of the book

As mentioned, the book is devoted to understanding how robotization in the short run, say the next 20 years, will influence the various manifestations of economic development: employment, per capita income, poverty and inequality, and economic growth. Automation is not a new thing but, as mentioned before, robots are autonomous machines which are aware of their environment and even capable of manipulating it and are therefore capable of functioning for long periods of time without human intervention. Thus, robotization is automation in a very new form. Given that robotization is not merely an intensification of previous automation but is qualitatively different, Chapter 3 investigates how robotization/AI has impacted production processes in various economic sectors, devoting attention to comparisons of human labour with robotic labour in terms of efficiency, and conducts some speculation regarding possible future impacts. Chapter 4 is devoted to a study of machine learning as this is a process through which robots can reprogramme themselves and adapt to the environment. An important question is whether this includes picking up soft and fuzzy skills

– including emotional and social skills – which robots are currently deficient in. In other words, the important question is: are there any areas which are beyond the reach of machine learning, given the current state of the science? Second, what are the implications of this for the future of human employment? The answers to these questions, based on a review of the work of scientists researching on machine learning, will have a lot of implications for the predictions and recommendations for policy made in this book.

Chapter 5 then reviews the empirical and theoretical work done to date on the economic impacts of robotization and marshals the lessons learnt and empirical evidence collected to provide a rich framework for analyzing how robotization will affect economic development, as characterized by the mentioned variables. Chapter 6 looks at various policy recommendations, which can help the human race take advantage of the possible benefits from robotization while enabling it to avoid its adverse impacts. This includes two types of policies: economic policy and those relating to skill and human capital formation, i.e., education. Chapter 7 differs from what was planned as it includes the impact of the Covid pandemic. But I have come to believe that the Covid pandemic might have a lot to do with how accepting the human race becomes of the adoption of AI in various walks of life. The influence of the Covid pandemic on the mentioned adoption and thereby on economic variables such as employment and income is therefore the subject matter of this chapter. Chapter 7 will conclude and provide a brief but succinct summary of (i) what the book has done to advance our understanding of the economic ramifications of the AI revolution as well as (ii) the value added by these advances.

Notes

1 As Adam Smith (1776) famously said in *The Wealth of Nations*, "Consumption is the sole end and purpose of all production". Even when workers put part of their incomes into savings, thus facilitating investment in return bearing projects, the purpose is to provide for future consumption.
2 Supporting evidence is provided in Nicholas Crafts (1999). Drawing on the Finnish experience, Mika Pantzar (1992) points to the significant increase in product variety in the 1970s.
3 This is a guesstimate based on data provided by the United Nations (website), according to which countries in Europe and America are often marked by small household sizes of less than three persons, whereas those in Asia and Africa often see average household sizes of more than five persons. The figure of 1.5 billion households can be obtained by dividing the world population of around 7.5 billion by 5, an approximation for the average household size in the world today.
4 Our discussion of the relationship between computer technology and unemployment is based largely on Ron Edwards (1987).
5 ThoughtCo. (website).
6 Swarthmore College (website).
7 Encyclopedia.com (website).
8 Ibid.
9 RobotShop.
10 Ibid.

11 WIRED (website)
12 RobotShop (website document)
13 Ibid.
14 Source: https://www.robotics.org/joseph-engelberger/unimate.cfm
15 Technopedia (Website).
16 Ibid.
17 USB (Universal Serial Bus) is the most popular instrument used to connect a computer to devices such as digital cameras, printers, scanners, and external hard drives.

3

ON HOW ROBOTIZATION HAS IMPACTED PRODUCTION PROCESSES AND THE RUDIMENTARY ECONOMIC IMPLICATIONS OF THESE IMPACTS

Developing a conceptual basis

In order to give bite to our discussion it is necessary to clarify the precise meaning of the term "robot" as it would be used in the rest of the book, though this concept was touched on briefly in Chapter 1. It is necessary to point out that all robots are machines though all machines are not robots. To be a robot a machine (i) has to be aware of its surroundings if such awareness adds value to the perception of its surroundings, and (ii) should be able to modify its behaviour, if necessary, according to changes in its environment if that environment is not totally unchangeable.

Take for example a helicopter. A helicopter should not land without human supervision: it has to be piloted. If it is not piloted, it can run into obstacles and its mission of transporting passengers can be short-circuited by collisions. A human pilot is therefore a necessary companion. A drone which is aware of obstacles in its immediate environment and can navigate around them is however a robot. It does not need any human companions to complete its task of moving from one place to another. Thus, operating a drone is less labour-intensive than piloting a helicopter, with the former a robot and the latter not one.

Let us take another machine, a conventional printer. It is not a robot because it is not self-sufficient: when out of paper it cannot reach out with an arm and fetch paper from a ream. It also does not possess the faculty of distinguishing between printable and non-printable paper. The function of feeding paper into a conventional printer, therefore, has to be performed by a human. If a printer is equipped with a mechanized arm capable of feeding it with paper when needed and distinguishing between printable and non-printable paper, it becomes a robot. A robot translates its awareness of the environment around it to function without a break for a significant period of time. If we extend this definition a

DOI: 10.4324/9780429340611-3

little bit every computer which is sensitive to the instructions from the user and the messages from the apparatus around it is also a robot.

Having clarified the meaning of a "robot", as it will be used in this book, let us examine the meaning of the term "robotization". A production process is said to be "robotized" if one or more robots are used in the production process. If as a result of robotization there are no humans involved in the production process, then it is said to be "fully robotized"; else it is said to be "partially robotized". A robotized production process thus is usually a substitute for another production process carried out with the help of non-robot machines and human players.

Thus, consider the non-robotized process of printing on paper where a human feeds printable paper into a conventional printer connected to a computer as well as sends signals to the printer through the computer to print designated pages in a document. This process can become partially robotized if the printer has an arm, assisted by robotic vision, such that it is able to reach out for printer paper and feed itself for printing. It can become fully robotized if the robotic arm is also equipped with devices which are capable of manipulating documents on the computer and managing the interface between it and the computer for the process of printing. In this book, I also use the term, "robotization" to refer to the enhancement of the role of artificial intelligence (AI) in the processes of consumption and production, involving one or more of the following: computers and similar devices, internet, and robots.

Next, I come to the term "production process" as this process involves the combination of "capital", defined as manmade means of production, and labour to produce output. Therefore, the nature of production processes in various sectors determines total employment in an economy. I define a production process as something which adds value to an input through the employment of capital and labour. For example, the process of stitching can add value to a piece of cloth by converting it into a shirt. Such a production process becomes partially robotized if one or more robots participate in it but humans continue to play a key and time-intensive role in the production process, possibly assisted by one or more non-robot machines. It becomes fully robotized if the participating robots perform various functions such that the conversion of a piece of cloth into a shirt is accomplished without the participation of a single human.

We can now easily see that partial or complete robotization has a direct impact of displacing humans from production processes. While partial robotization reduces the amount of labour per unit of output (labour intensity of output), it will only be adopted to replace a production process which does not involve robotization if it brings down costs per unit of output, a phenomenon accompanied by price reduction and increase in output/sales. One major exception to this conclusion is the agricultural sector: when farmers endowed with a given piece of land can use robots to produce a higher output at higher cost per unit of output, robotization might still be more profitable than conventional production – because of a lower profit margin on each unit being overwhelmed by greater sales – and an option which farmers might choose.

Because labour employed is the product of the scale of output and labour intensity of output, the impacts of increases in output and decrease in labour intensity on employment are in opposing directions. Both increase and decrease in employment in the sector of focus are therefore possible through partial robotization. With full robotization the amount of human labour used in the production process becomes minimal and therefore its introduction by an enterprise to replace partially robotized or un-robotized production processes will inevitably bring down the direct use of labour in production.

However, note that both fully and partially robotized production can have positive indirect impacts on human employment. Recollect the increase in the scale of output following a switch from conventional production. This implies that more humans will be needed to transport output to warehouses from the production facility, and from warehouses to markets as drivers of mechanized trucks. Similarly, more warehousing capacity would be needed, and more humans would be needed to man these warehouses. And as mentioned in the last chapter, saved income, because of price fall originating from the decrease in cost of production facilitated by robotization, can stimulate demand for various industries, thus contributing to employment increase.

It is important to point out though that the negative direct effects on employment arising from the diminished labour intensity associated with partial or full robotization might come into play long before the positive indirect impacts are felt. Consider for example the use of robots to manufacture smartphones (Market Research Future: website). Between 2017 and 2023, robotized manufacture of smartphones is expected to grow fivefold to a value of $4 billion. Given that this technology is available and enables production at lower cost and higher speed, this is to be expected. But it will surely hit the direct employment of humans in smartphone manufacture. Gradually however, the positive effects will be felt: less affluent people in developing countries such as small farmers and those employed in the informal sector will become first time owners of smartphones and be able to access information regarding markets and economic opportunities as well as free entertainment, with salutary implications for economic activity and employment generation; existing owners of smartphones will be able to go in for faster replacement of instruments as well as be able to spend more on other products, including ones which are labour-intensive. But while it is true that the positive indirect impacts on employment might ultimately overwhelm the initial negative impact, the initial job destruction will probably render people jobless for significant periods of time. Therefore, there is a role for policymakers to step in and manage the negative fallouts of robotization, irrespective of whether the net impact is positive. There are significant psychological ramifications of unemployment: social safety nets can provide succour to those render unemployed and eventually enable them to rejoin the list of contributors to the economy as well as providers of the means for their own subsistence.

The impact of AI (robotization) on production processes

Futurists sounding alarms about the job displacing effect of robotization has been seen in many quarters as a story of the child repeatedly crying "wolf". In the 1960s, a report called *The Triple Revolution* was written by a group of prominent academics, journalists, and technologists referred to as the *Ad Hoc Committee on the Triple Revolution* (1964). This included Nobel laureate chemist Linus Pauling and economist Gunnar Myrdal (who was to get the Nobel Prize in the 1970s). A large part of this report focused on cybernation, now referred to as artificial intelligence. The report claimed that the world was on the brink of having systems of machines producing potentially unlimited output with limited human help. The report warned that this system would lead to massive unemployment, soaring inequality, and falling demand for goods and services. It recommended the institution of a guaranteed minimum income programme.

However, in the period immediately following the report, the doom and gloom forecast by this distinguished committee was never realized. For a good two decades after this report came out, humans and machines remained largely complements: it was not possible to operate machines for significant lengths of time without the constant supervision and interference of humans. While it was true that labour intensity of production was decreasing, the increase in the scale of production of existing product varieties and increase in product variety implied that there was no discernible increase in unemployment. Until 1969, the unemployment rate in the United States never went above 7% in even recessionary periods and much of the growth in productivity was matched by higher real wages. The march of technology decreased the demand for blue-collar or manual work despite the increase in the scale and scope of production, facilitated partially by the demand side and increase in per capita income, which however ensured that white-collar work increased.

However, Ford (2015) points out that after 1973, the United States entered into a phase of continuously declining average real wage (i.e., the money wage adjusted for increase in prices, which gives us a measure of the purchasing power of a typical worker) that was obviously a consequence of greater abundance of labour relative to demand. In four decades, the average real wage of non-supervisory workers plummeted to 13%. This was matched by the income distribution becoming more skewed with time: the ratio of the median income to per capita income, which was constant in 1949–1973, declined in the next 30 years by 32%.[1] This happened because blue-collar workers saw their share in national income drop very fast, a phenomenon which was associated with the share of wage or salary earners of which the group of blue-collar workers is only a part, declining from 65% to 58%. Note that a significant decline in the share of labour in national income has been experienced by most countries in the world in recent times.

Going back to the case of the United States, and referring to Ford (2015) again, it should be noted that in the first decade of the 21st century, the net number of jobs created in the entire economy was practically zero. This was of course attributable to the job losses during the Global Recession but even if we use the average annual job creation rate for the pre-recession period of 2000–2007 as the average for the entire decade, the increase in employment in the entire decade would have been 8%, less than half of that of the 1980s and 1990s. Second, as the economy emerged out of the recession there was obviously a tendency to increase output through available new capital-intensive technologies based on automation rather than by creating jobs, to the extent that the new machines or robots were easier to manage than humans and more cost-efficient in regard to production of output. In other words, what emerged was a tendency to transition to a lower labour intensity of output as growth started reviving after the recession.

Note that total employment is given by the product of labour intensity of output and output. Thus, growth rate of total employment is given by the sum of the (a) growth rate of labour intensity of real GDP and (b) growth rate of real GDP. Since the former has been negative due to robotization and other automation, this has tended to reduce the growth of jobs. Further note that over time as per capita income grows as a result of the introduction of less labour-intensive and more advanced technologies and costs and prices plummet while profits increase, the share of labour in national income tends to go down, given decreased labour demand reducing wages and employment. Data shows that this is indeed the case. As the providers of labour services are marked by a higher propensity to consume than owners of capital owing to their lower incomes, the expansion of markets accompanying economic growth is slowed down. This in turn can reduce the rate of economic growth.

The workers with the highest propensity to consume are often the blue-collar workers. The mentioned decline in real wage and job creation meant that around the end of the 20th century and the beginning of the 21st century, the share of the blue-collar workers in total national income in the United States reached unprecedented low levels. This was associated with more than 50% of the increase in real GDP in the period 1993–2010 going to the top 1% of the income ordered population.

In what follows I shall look at the impact of robotization on production processes in agriculture, industry, and the service sector, identifying sub-sectors in which labour has been substituted by robots and others in which the reliance on human labour has not been greatly diminished. The potential for robotization in various sectors is also evaluated.

The impact of robotization on agricultural production

Automation of developed country agriculture has been going on for well over a century and has displaced huge amounts of labour. Between the late 19th century and 2000 the proportion of the workforce that was employed in developed country agriculture dropped from 50% to 2%. The U.K. suffered a similar drop:

in the 19th century, 22% of total employment was in agriculture and fishery, whereas today that proportion is less than 1%. Examples of mechanized processes that have been in use for a long time are robotic milking and automated slaughtering. However, some jobs have continued to be done by humans.[2]

Developed countries are now witnessing a new wave of labour displacement through the introduction of robots, which unlike the machines introduced in the previous wave, can often function for long periods of time without human interference. In this second wave there is a tendency to turn over even the more nuanced activities such as picking of fruits and flowers, requiring judgement about colour and texture to detect ripeness, and pruning of vines from humans to robots.[3] Human judgement here is substituted by a combination of machine vision technology[4] and algorithms. In addition, weed picking, phenotyping, or the use of knowledge about DNA to predict the characteristics of plants and therefore form ideas about yield and quality of produce is being done by robots/AI.

Common sense dictates that "enhancement of AI" or "robotization of an industry" leads to scale and substitution effects. First, as pointed out, robotization leads to human beings in a production process being displaced by robots. This can be termed as the "substitution effect". For example, replacement of conventional airplanes by drones reduces the demand for human pilots. Replacement of a commonplace printer by another with a mechanical arm, as explained in the last section, would make humans redundant in regard to the process of printing. In agriculture, the introduction of robots in various agricultural operations such as harvesting and weeding will displace humans from employment. This would have a negative impact on total employment of humans in agriculture. But at the same time there is a scale effect: higher cost efficiency of robotized production will not only induce a switch to it from conventionally mechanized production but also production of a higher amount of output. This would imply that the marketing and transportation of output would have to take place on a larger scale, thus generating additional opportunities for the employment of humans. In due course of time, however, marketing and transportation might become fully robotized. The discussion of robotization of agriculture in this book will cover both substitution and scale effects.

Why is robotization needed in agriculture and what determines its adoption?

The use of robots in agricultural operations will in all probability usher in an era of *precision agriculture* (this term is discussed and defined below), which will increase agricultural yields and generate cost reductions, thus helping to feed a rapidly expanding and prospering human population apart from enhancing farm incomes. I elaborate on this argument below.

Between 2019 and 2050 world population is expected to increase from 7.71 billion to 9.74 billion according to the medium variant of the *World Population Prospects 2019* (United Nations, 2019) which assumes that fertility will decline in countries where large families are present and increase in countries where it is currently on average at replacement levels of two live births. Thus, the total increase is a very large 2.03 billion that amounts to 26.2% of the world population in 2019. This increase of more than 2 billion will be accounted for by population growth in various large countries (I define a large country as one which will figure in the ten most populous countries in 2050 according to the mentioned projections) to a very significant extent as follows: India, which has been experiencing rapid economic growth (i.e., growth of gross domestic product) over the last three decades and is expected to do so in the future, will see a population increase of 273 million (13.50% of the total global increase); Nigeria, which is currently experiencing slow economic growth but might return to the fast growth trajectory witnessed in 2000–2014 through structural reforms prescribed by the World Bank (World Bank, 2020[a]), will witness a doubling of its population through an increase of 200 million (9.9%); the Democratic Republic of the Congo, which has experienced moderate annual economic growth of 3.4%–5.8% in the recent past (African Development Bank Group, 2020), will undergo a trebling of its population through an increase of 127 million (6.28%); Pakistan, which experienced moderate to high annual economic growth of 1%–5.8% in 2015–2019 (World Bank, 2020[b]), will witness a population increase of 121 million (5.98%); and Ethiopia, which experienced high average annual economic growth of 9.9% (World Bank, 2020[c]) in the decade ending 2017–2018, will experience a population increase of 93 million (4.60%); together these five currently rapid or moderately growing countries, whose per capita incomes are very likely to increase by a large amount in 2019–2050 given their current low levels and demonstrated capacity to grow, were home to 25.45% of the world population in 2019 but will account for 40.26% of the total projected increase of 203 billion in 2019–2050. Four out of the other five large countries – United States, Indonesia, Brazil, and Bangladesh – accounting for 12.63% of the world population in 2019 will witness an increase of 158 million in their population, thus contributing 7.78% of the projected increase. Two of these – Bangladesh and Indonesia – are rapidly growing countries with levels of per capita income which are still low by world standards and therefore should see these incomes increase very significantly over the next 30 years. Finally, China, with a population of 1434 million or as much as 18.59% of the world population in 2019, is the only large country which will experience a population decrease (of 32 million) during the period 2019–2050. It is still a middle-income country and therefore its rapid economic growth will in all likelihood continue over the next three decades and result in significant growth of per capita income, especially because it has been able to contain its population growth.

More than 50% of the population increase in 2019–2050 will come from the countries of the world which do not fall into the category of large countries as defined above. Let us refer to these as small countries. Small sub-Saharan African countries, i.e., countries other than Nigeria, the Democratic Republic of the Congo, and Ethiopia, currently home to less than 9% of the world population, will account for a population increase of 632 million (see Table 3.1) or 28.57% of the total global increase of 2.03 billion. All these countries have extremely low levels of per capita income with enough scope for growth in per capita income.

To summarize, 83% of the projected global population increase of 2.02 billion in 2019–2050 will occur in developing countries with low or middle levels of income and much catching up to do with the developed countries. Table 3.1 identifies the prime movers among developing countries in regard to population increase in 2019–2050: the large and small countries of sub-Saharan Africa, home to less than 14% of the world population in 2019 and accounting for as much as 51.1% of the global population increase in 2019–2050; and the two countries of India and Pakistan, home to 20.5% of the world population in 2019 and accounting for a matching 19.5% of the global population increase in the mentioned period. The prime movers, all developing countries, thus account for about 34% of the world population in 2019 and around 70.6% of the global population increase. Developing countries other than these prime movers will account for 12.4% of the global population increase. Note that many countries among the prime movers, including all the large ones, and other developing countries such as Indonesia and Bangladesh have already demonstrated a capacity to register rates of economic growth far in excess of population growth and thereby enhance their per capita incomes significantly. Given that their per capita incomes are still very low, (a) there is room for further significant increases in per capita incomes which is likely to be exploited in the future, given their demonstrated potential for economic growth; and (b) such increases would in all probability be associated with significant increases in per capita demand for food, chiefly non-staples and proteins. The mentioned large population growth in these countries will obviously be another major source of increase in the

TABLE 3.1 A Summary of Projected World Population Increase in 2019–2050 with Emphasis on the Contributions of Prime Movers

	Population in Million		Population as % of World Population		Increase in Population during 2019–2050 (POP CHANGE)	POP CHANGE as a Percentage of Increase in World Population
Year	2019	2050	2019	2050		
World	7713	9735	100.0		2022	100
Nigeria, Democratic Republic of the Congo, and Ethiopia (NCE)	380	800	4.9	8.2	420	20.8
Sub-Saharan Africa exclusive of NCE	686	1318	8.9	13.5	632	31.3
India and Pakistan (IP)	1583	1977	20.5	20.3	394	19.5
Total for SSA + IP	2649	4095	34.3	42	1446	71.5

Data source: *World Population Prospects 2019* (United Nations, 2019).

demand for food. Note that though China – with the mentioned large chunk of the global population – will see a small decrease in population as mentioned, significant increases in per capita income in the near future, expected on the basis of current trends and level of per capita income, will imply that it too will continue to be a major source of increase in world demand for food.

It should also be noted that the ability to meet the increase in demand for food resulting from economic growth and population growth has implications for economic growth itself. The inability of the agricultural sector to increase the supply of food to an extent that matches the increase in the number of mouths to feed and per capita income will result in food shortages, food inflation, and increase in the incidence of malnutrition, with the last phenomenon occurring mainly in developing countries characterized by a significant proportion of population living below the poverty line. With both savings and human capital formation threatened, growth in per capita income and associated poverty alleviation will be checked. Thus, it is very essential that measures are taken to enhance the supply of food through (a) yield increases brought about by the adoption of new technologies and innovations in the cultivation of land and (b) increase in effective acreage, where possible, through multiple cropping and measures to cultivate more land.

Data on demand for food has invariably been found to be consistent with Engel's law (Engel, 1857): as per capita income increases, the proportion of income spent on food decreases. This makes quite a lot of sense: food such as staples are necessary for providing energy requirements that are the basis of life but there is no need to consume greater amounts of these staples as a human ascends to higher levels of income, especially when skilled labour is usually associated with lower expenditures of human energy. Investigations however reveal that the absolute level of per capita expenditure on food increases with income, i.e., the decrease in proportion of income spent on food is small enough for the product of such proportion and income to increase. Furthermore, other findings reveal that a growth of per capita income is associated with a tendency for higher purchase of quality foods, such as vegetables and proteins (see Clements and Si, 2018). In other words, there is a tendency for diet diversity to increase with per capita income. We can expect that in the next 30 years, the rate of increase in demand for staples such as wheat and rice, though significant, will be lower than that for fruits and vegetables as well as dairy and meat: at very low levels of personal income, consumers predominantly demand staples and this demand increases with increase in personal income, reaching a plateau at a low enough level of income (thus, those in the upper middle class do not consume a higher amount of staples than those in the lower middle class); in regard to non-staples such as fruits and vegetables or many proteins, consumption becomes significant or noticeable only after the consumer has reached a high enough level of income, and rises with further increases in income, reaching a plateau at a level of income much higher than the level associated with the start of the plateau for staples. Most developing countries have now attained a position where per capita demand for staples has reached or neared a plateau; aggregate demand for staples will now be driven upwards predominantly by population growth. In regard to non-staples, the upward drive will be powered by an increase in per capita income as well as population and therefore will be characterized by a much higher rate of growth.

Note that technologies will be introduced to increase the per capita supply of food for a growing population if such technologies increase the income per hectare of agricultural producers. I argue below, in a detailed manner, exactly how the introduction of robots and artificial intelligence can enhance the incomes of agricultural producers through yield increases and cost-reducing tendencies, as well as reduction in risks emanating from possible shortages of human labour and the inability of human labour to work very long shifts when there is a requirement – for example, a need to harvest the crop very quickly because of expected bad weather. The argument leads to the following broad conclusion: robotization in agriculture can result in win–win solutions – more abundant food at a lower price and greater farm incomes with smaller year-to-year variation due to diminished risk.

There are three mechanisms through which agricultural robots can enhance yield, add value, and reduce cost. Introduction of robots is labour-saving: some or all of hired labour might be set free leading to a reduction in the wage bill. Second, introduction of robots results in more economical and optimal use of other variable inputs such as fertilizers, pesticides, herbicides, and water through *precision farming*: it is possible to precisely target the plants with required inputs and thus wastage and excess is almost eliminated. Third, robots can supervise crop health and hand over processed data in this regard to the farmers for steps that can optimize crop yield and quality. Thus, when a farming household introduces robots into cultivation it gains, as a result of higher revenue caused by higher yields, a reduction in the wage bill for hired labour and a reduction in expenditure on the mentioned variable inputs. Expenditure on electric power may increase. Total capital expenditure increases or decreases according to whether new capital expenditure on the services of robots gets offset by reduced capital expenditure on the services of non-robot machinery. Enhanced capital expenditure is a strong possibility. Farmers also might not let go of conventional machines such as old-style tractors if they are in good enough shape to render a significant amount of service in the future.

Note that economic profits are given by total revenue (the product of farm size, yield, and price of unit produce) minus labour costs minus capital costs minus expenditure on variable inputs such as fertilizers, pesticides, herbicides, water, and power. Importantly, economic profits are calculated by imputing a value to employed family labour at the going market wage and considering this as part of labour cost. The total labour costs are thus calculated as the sum of the total wage bill (money actually spent on hired labour) and the imputed value of family labour, defined as the product of the farm wage rate and the quantity of family labour used. For the sake of simplicity and without loss of generality, let us assume that no member of the farm household is employed outside the farm prior to robotization. The economic profits plus the imputed wage income of the entire family labour constitutes the true total mixed income of the farm household. Further let us assume, again for the sake of simplicity and without loss of much generality, that labour displaced by robotization is less than the amount of hired labour and labour displacement occurs purely through the laying off of hired labour. Thus, robotization produces no change in the amount of family labour used by the household and therefore in the imputed value of family wage income.

Economic profits tend to increase as a result of robotization because of the following tendencies: higher yields; lower use of variable inputs such as fertilizers, pesticides, herbicides, and water; lower employment of hired labour; and lower crop losses because of better monitoring of crop health. Increase in expenditure on electric power and potential increase in capital expenditure associated with robotization tend to produce a contractionary impact on economic profits. Economic profits actually increase when the former expansionary tendencies overwhelm the latter contractionary tendencies, a strong possibility. When economic profits increase, the total mixed income of the farm household gets enhanced as the imputed value of farm wage income remains unchanged. Farmers will actually make a switch to a robotized technology from a conventional technology if and only if they anticipate an increase in economic profits due to robotization. As robot production benefits from learning-by-doing and economies of scale, the required capital expenditure on robots by farmers considering robotization would go down. Thus, the mentioned switch would become profitable for the very farmers who had earlier deemed it as not profitable.

It is also possible that robotization may make a certain amount of family labour and the entire hired labour redundant. Assume that at any point of time, the wage rate used to impute/calculate the income of a unit of family labour is the market rate and is the same irrespective of whether that unit is employed on the farm or outside. The choice between the two options of going in for robotization and sticking to the conventional production technology therefore has no implications for the total value of family labour income. In this case, mixed income of the farm household will always get enhanced by robotization augmenting economic profits. Thus, as long as robotization enhances economic profits, the farm household will switch to robotization. However, as robotization proceeds, job displacement of labour should lead to an excess supply of labour and bring the wage rate down. Thus, the desirability of robots as substitutes for human labour goes down with the latter becoming cheaper.

Consider a scenario in which it takes different amounts of time for farms to get informed adequately about the robotized technology and react to it. Thus, the process of robotization unfolds over the concerned region over a period of time of considerable length. As robotization unfolds and sweeps across the region it would be associated with the wage rate for human labour decreasing. Robotization would come to a stop when households who have recently received information about the robotized technology and are about to act on it find that the wage rate has become low enough for a switch in technology to be no longer profitable. Here it is possible that increasing adoption of the robotized technology and a decline in wage over time is such that the adoption remains profitable till every farm household has switched over to the new technology, an outcome made likely if this technology is vastly superior to the traditional technology and/or the initial supply of labour is low resulting in a high wage rate and greater gains from substituting human labour with its robotic substitute. Populous developing countries such as India and Bangladesh, characterized by low wages by international standards, will obviously not be characterized by this outcome: only a fraction of the total population of farm households, very likely those who are educated and therefore tuned into information about available technologies, might think of adopting robots. Note that our discussion has implicitly considered farm households to be homogeneous except with regard to their "reaction times". But in reality, there might be heterogeneity in respect to other criteria such as size, again correlated with farm income, and the average level of education within farm households. Large farmers, given their higher level of education and sensing an opportunity for furthering their self-interest, might be more alert regarding information about new technologies, such as the robotized one, which display greater economies of scale than the traditional one. It is these farmers who might react to the information about the emergence of the robotized technology and adopt it.

The crucial assumption here is that if family labour is displaced from the family farm by robotization then it will surely find employment outside. But in reality, the likelihood of not finding a job might be significant and increasing in the extent to which the robotized technology has already been adopted by farms in the region. This is especially true when wage rate for labour is sticky and only offers subsistence, with employers not willing to pay a lower wage, fearing an adverse impact on the efficiency of labour, or potential employees not interested in a wage that does not pay even for subsistence. Thus, the farm household has to weigh the gains from enhanced profits induced by robotization against the risk of some of the displaced family labour remaining unemployed and not earning any income. As robotization proceeds, this risk increases, ultimately causing farms at the margin to stick to the traditional technology; there is a cessation of robotization. Alternatively, if the robotized technology is far superior to the traditional one, the enhanced risk will continue to be overwhelmed by the boost in profitability that robotization facilitates till the robotized technology is adopted by the entire region.

Clearly the gap in profitability between the robotized technology and the new technology is crucial; if large enough, it can ensure that all farm households will adopt the robotized technology. However, again, this is unlikely to be the case when the level of the wage is low or just able to provide the means of subsistence. In such countries, we would expect a fraction, far less than one, of the farm households to adopt the robotized technology.

To repeat, I have assumed here that the time taken by a farm household to react to the arrival of the robotized technology varies across households; the farm households which respond faster to the opportunities for robotization, say because of faster information flows about the robotized technology or greater flexibility in thinking, are faced with better prospects for reemployment of family labour outside the farm than those who respond more slowly. This is because the time elapsing between the two types of households being informed is marked by substitution of human labour by robots.

In the next sub-section I shall consider the case where the time taken to react to opportunities for robotization is the same for all households. Note that these discussions apply more to developing countries which currently have a very large proportion of the labour force employed in agriculture. On the other hand, the proportion is so small in developed countries that a decrease in agricultural employment due to robotization will not be a very significant phenomenon, and therefore not a bother for farm households or policymakers.

As mentioned before, the use of robots in agriculture has ushered in what is known as *precision agriculture*, which basically means, among other things, that inputs such as fertilizers, pesticides, and herbicides will be used in the precise quantities required; ploughing will be done with just the required amount of force; plant health will be monitored by ascertaining the health needs of each plant and catering to these needs, an approach which yields a positive return because health needs vary across plants; and loss of produce during harvesting will be minimized. Precision agriculture is associated with the use of cutting-edge robots to perform operations that have been handled until now almost entirely by humans: plucking of fruits and flowers and harvesting of vegetables; pruning and thinning of plants; removal of weeds or spraying of weeds with herbicides after identification; rearing of plants in greenhouses where humidity, temperature, and light can be controlled to optimize growth; and aerial spraying of seeds as well as supervision of plant growth. In what follows I take a detailed look at the mechanisms through which precision agriculture can enhance crop yield and reduce farming costs by considering actual examples of robots. Our discussion is based on Gosset (2019), Alexander (2018), and other texts such as Pinduoduo (March, 2021).

Consider the use of a large conventional tractor to plough the soil and make it ready for supporting a crop. The primary purpose of ploughing is to turn over the uppermost soil and bring the nutrients that help crop growth to the surface while burying weeds. The mentioned large tractors can kill worms and damage fungi and microbes, thus negatively impacting soil health. Small autonomous and smart tractors — such as those manufactured by the Brazilian equipment supplier, *Stara* — which have been introduced recently are not associated with these negative impacts. Besides, these are like offices on wheels, which help to monitor the soil through sensors and collect data that enable the provision of optimal doses of water and fertilizers to plants as well as determination of the best time for their harvest.

Very small robots, such as those manufactured by the U K.-based *Small Robot Company*, monitor soil quality and plant health and maturity during the growing season as well as look for signs and severity of disease so that interventions can be made at the right time. Some companies have started manufacturing unmanned aerial vehicles for farmers. These provide an aerial view of the crop which give the farmers a good idea of crop health as well as weed growth, irrigation layout, and insect issues. These machines can plant seeds accurately as well as come up with an estimate regarding the amount of pesticide needed.

In regard to robots which monitor plant health, consider *Scout*, a "drone" manufactured by *American Robotics*, which lives inside a weatherproofed box where it can self-charge and process the data it collects. When a flight over the fields is needed to examine the crop, the box opens and the drone lifts off for an aerial survey powered by artificial intelligence. During such flights *Scout* gathers crop stress data to guide the decisions of farmers for the practice of the so-called precision agriculture.

Robots can not only collect information on plant and soil health and threats to such health during the growing season but also make the necessary interventions. For example, consider weeding robots which are gradually becoming popular. Robots have been developed which can move through the rows of plants to deal with any weeds that might have sprung up and spray the weeds in a targeted manner with herbicides. An example of a weeding robot is *Ecorobotix*, a drone which taps solar power to work all day and uses its complex camera system to identify and spray weeds. It uses 90% less herbicide than human workers which lowers the cost of agricultural production as well as the chemical content in agricultural produce, thus potentially generating value for consumers through lower prices as well as healthier foods. Such robots effectively supplement the weeding by larger robots before the planting of the crop, using blades and finger-like devices. The practice of robotized removal of weeds is not only replacing the difficult and time-consuming practice of human weeding but also helping to check excessive application of herbicides by farmers which not only induces resistance in weeds but enhances the concentration of chemicals in plants and therefore food. Precision weeding, the robot-powered removal of weeds during soil preparation and the targeted spraying of weeds by robots during the growing season, also has the potential to enhance yield as weeds compete with the crop for soil nutrients and pesticides.

There is another channel through which precision agriculture generates beneficial implications for the yield of fruit crops such as almonds, apples, blueberries, avocadoes, potatoes, onions, and beet which rely upon insects to produce fruit or seeds. There has been a decline in the insect population due to disease, climate change, and the use of pesticides which does not augur well for these fruit crops. A drone devised by the American firm, *Dropcopter*, can spray trees with pollen. The pollen sticks to the trees and are moved by the crowd of feeding insects to pollinate flowers. The targeted spraying of pollen helps the reduced population of insects to satisfactorily facilitate the job of pollination.

This system is better than that used in recent times to cope with the reduction in numbers of pollinators – costly transportation of swarms of honeybees to carry out pollination, a process which works to the detriment of local insects. Besides, the number of honeybees has been plummeting due to disease, a development which means that their mentioned use for pollination might be unsustainable.

Research commissioned by *Dropcopter* suggests that this method of pollen spraying by drones can increase yields by 25%–50%: unlike honeybees, with their scattergun flight paths, it can ensure that all trees receive pollen; thus, there is no need for the farmer to plant alternative rows of fruit-producing and pollinating trees but instead she can just concentrate on fruit-producing trees.

Consider the harvesting robot which is a major innovation as unlike the harvesting human it can work 24-hour shifts. This helps to complete the job of harvesting in a shorter period of time and thus minimize loss of harvest due to weather fluctuations. Consider the advanced strawberry-harvesting robot called *Berry 5* which uses various robotic components to grab the leaf, pluck the berry, and pack it. Computer vision helps *Berry 5* to distinguish between ripe and non-ripe berries before plucking. The *Berry 5* is faster than a human labourer as it can pick a plant in 8 seconds and shift to the next in 1.5 seconds. Similarly, the *Energid Citrus Picking System* is a robotic system that can pick a citrus fruit every two or three seconds. *Abundant Robotics* has introduced an apple-gulping robot in California. Sophisticated computer vision enables the robot to gulp up ripe apples while bypassing the unripe ones. The algorithm used by it to pluck the ripe fruit can be updated on the basis of farmer feedback about how good a job it is doing in plucking on the basis of ripeness. The *Vege-bot* is a recent robotic innovation developed to pick lettuce which until very recently has been considered robot-resistant because of its fragility and close proximity to the ground. The robot uses a camera and a machine learning algorithm to identify unripe and diseased lettuce. Instead of human hands the *Vege-bot* uses a blade assisted by the camera to harvest lettuce.

Pruning of plants such as vines can also be a labour-intensive operation. *Vision Robotics* has come up with a robotic system that uses artificial intelligence and machine vision to identify plants for thinning and then uses a robotic arm to undertake the actual operation of thinning/pruning.

A few robotic greenhouses are also coming up. Plants are grown in an indoor environment and draw sustenance from nutrient-rich water instead of soil. Artificial lighting substitutes for sunlight and there exist systems for controlling temperature and humidity and providing filtration. Robots supervise the growth of plants and are also involved in physical tasks such as the transportation of plants as well as their planting in optimal patterns. It is estimated that these greenhouses require 90%–95% less water than soil-based agriculture to generate an equivalent yield. Moreover, the use of controlled indoor environments implies that pesticides, harmful for consumers, are not used.

The above discussion has shown how robots, an integral part of precision agriculture, can really make a difference to agricultural yield and reduce cost through greater economy in the use of variable inputs such as labour, herbicides, pesticides, and water. This analysis is consistent with the prediction of *International Food Policy Research Institute*: precision farming techniques and new technologies could help to increase crop yields by up to 67%.

Robotization and the agricultural economy: detailed discussion on implications for wage and human employment

Until now we have looked at how robots can increase yield and reduce costs, thereby motivating farmers to robotize when appropriate robotic technologies become available. But this displacement of labour at the level of the farm does surely have implications, when aggregated across farms, for market demand for labour, the wage rate, and the profitability of farming operations. There has been some discussion on this topic in the previous sub-section to motivate appreciation of the process underlying the determination of the extent of robotization in a scenario in which time taken to respond to the availability of the robotized technology varies across farms.

The Appendix to this chapter examines the implications of the process of robotization for wage rate and human employment through a stylized model which considers the time taken to respond to the availability of robotic technology to be uniform across farms and instead focuses on the implications of another characteristic of robots – durability, i.e., the farmer has a choice between hiring the services of human labour or buying a durable means of production, the robot, which can provide the same services, but over multiple periods.

Consider a scenario, a close approximation of reality, in which robots cannot be hired but have to be bought and in which there is no second-hand market for robots. Instead, when farmers lay off some of the human labour employed by them to buy robots, they in fact become committed to using these robots for their entire life of T periods: as soon as the farmer buys a robot, she effectively becomes committed to incurring a cost per period, for T periods, equal to an amount c which equals price of the robot plus interest income flows sacrificed in buying the robot divided by T. For example, consider a robot whose price is $100 and which is capable of providing services for ten years. If the farmer buys the robot, she foregoes an interest income of $10 per year if the rate of interest is 10%. At the same time, she loses 10% of the value of the robot through depreciation if we assume that yearly depreciation accounts for a constant proportion of this value which reaches 0 after ten years. In terms of today's dollars therefore she loses $20 every year due to the purchase of the robot. Thus, the value of c is $20.

I assume that when the robotic technology is introduced, it involves displacing an amount of human labour in the production process with the same amount of robotic labour[5] and is able to bring about an increase in yield. The farmer however has the freedom to choose the extent of robotization, given that cultivation can be broken down into various operations such as tilling, weeding, planting, supervision of plant growth, harvesting, etc. An approximation is the farmer being able to choose the extent of robotization, a variable which can take any value from 0 to 1 ranging from no robotization (0) to full robotization (1), with the addition to yield caused by further robotization being smaller at a higher level of robotization. The last assumption is a product of common sense: if the robotization of an operation enhances yield by a larger amount than robotization of another operation then the robotization of the former operation precedes robotization of the latter if the farmer is rational.

The introduction of the robotic technology gives the farmer an opportunity to increase yield through robotization which involves substituting robots (robotic labour) for labour and incurring a cost proportional to the difference between c and the wage rate, the price of human labour. For the sake of simplicity, we neglect non-labour variable inputs. If this difference is negative, then robotization increases yield as well as reduces expenditure on inputs; the farmer therefore goes in for full robotization. If the difference is positive but very small, then it is possible that further robotization brings about an increase in yield which more than compensates for the increased expenditure on inputs at every level of robotization; the farmer again goes in for full robotization. A higher difference brings about partial robotization, with further increase in robotization not bringing about an increase in yield that is large enough to compensate for the increase in expenditure on inputs. When the difference is very large, which could be because human labour is plentiful and therefore the wage rate is low, the farmer chooses not to robotize as the enhancement of yield due to robotization can never compensate for the increase in expenditure on inputs. The farmer has no option but to wait for better and cheaper technology – a lower level of c and robots that can bring about larger increases in yield – so that robotization becomes profitable.

If the introduction of the robotic technology brings about partial robotization at the prevailing market wage it is associated with a dip in the quantity of labour demanded and therefore an excess supply of labour. As the sudden abundance of human labour drives market wage down, the farmer does not respond by increasing the quantity of labour demanded: having committed to practically pay for the services of robots over multiple periods at a price of c per period per robot by buying these, he obviously is better off using the services of these robots at no additional cost instead of substituting these by hiring additional services of human labour at an additional cost. Thus, the elimination of excess supply through wage reductions takes place through the sole medium of contraction in the quantity supplied of human labour. This enhances the magnitude of reduction in market wage which can be quite significant.

Consider an alternative to ownership of robots: the hiring of services of robots on a periodic basis. In this case, the excess supply of labour caused by the intro-duction of robots leads to a reduction in market wage which in turn eliminates this excess supply through both increase in quantity demanded as well as decrease in quantity supplied of human labour. This results in a reduction of wage that is much lower than that experienced for the case of ownership of robots.

Recent developments in regard to robotization in developing countries

While agriculture in developing countries continues to be much less automated than in developed countries presumably because of lack of purchasing power to buy robots and small size of agricultural holdings making production through robots unviable, policy changes seem to indicate that these countries might be on the brink of more job displacing automation and even robotization.

For example, in India, the Model Agricultural Land Leasing Act, 2016 (see Mani, 2016) has been drawn up to give a push through official legislation to, among others, contract farming. Furthermore, the Agricultural Produce Market Committee Act, 2003 (see Singh, 2015) now officially recognizes contract farming and facilitates dispute resolution in this regard, thus alleviating risks and enhancing engagement in agricultural contracts. This is paving the way for effective consolidation of hitherto small agricultural holdings. Automation and robotization might be consequences of the increased incidence of agricultural contracts: given that yield-improving robotized technology is available and more profitable than conventional technology on farms exceeding a certain critical size, agricultural contracts can be used to consolidate farm holdings and enhance profits; the increase in profits can be considered a surplus which can be shared among the various contracting parties, the involved farmers, and a firm which can provide crucial inputs and bear some of the risks.

Though the direct impact of such introduction of AI on employment opportunities in farming might be negative, the increase in agricultural production facilitated by AI would be associated with an increase in the level of employment in marketing and transportation of agricultural produce. This increase might however be reversed in the longer run because of automation of transportation and marketing.

To end this section, I look at certain digital initiatives taken in Indian agriculture, with digitisation being used to ensure greater transparency in the agricultural economy (Shrivastav, 2021). According to a report of NITI Aayog, which shapes or proposes policy incentives in the Indian economy, agriculture must grow at 4% or higher. This highlights the obvious fact that high growth in the entire economy must be based on a certain minimum amount of growth in agriculture. But this calls for more transparency in the functioning of agriculture: technology can be used for sensor-assisted soil assessment; modern family methods can be used for temporal and spatial vulnerability within plots of land; and digital technologies can be used for reducing waste generated in agriculture. Retail markets are also using the benefits of digital technology to augment the benefits of farmers, until now the subject of exploitation of middlemen.

The impact of robotization on manufacturing

It is seen that robots have a role to play in large-scale manufacturing: first, precisely timed movements can help them function as a team even without vision;

second, in case robots function in an environment affected significantly by humans which is therefore more uncertain, those with three-dimensional vision can function as a team without outside interventions. Such robots are also easier to train than humans; they can also be more easily managed and are quite amenable to working in 24-hour shifts unlike humans.

I now look at the impact of robotization and AI on manufacturing. As mentioned, routine operations in manufacturing are being carried out by two types of industrial robots: those with three-dimensional vision and others with precisely timed movements but without the mentioned type of vision. The second group of robots can, as mentioned, operate in an environment characterized by perfect certainty i.e., that characterized by entirely predictable events made out of their own precisely timed movements. Both groups of robots are being used in assembly line work for manufacturing or in warehouses which provide for efficient retailing. The efficiency of these robots has been increasing over time, so much so that it will definitely surpass that of humans in shifting, packing, and assembling objects.

The employment of such industrial robots has been growing over time. To illustrate, global shipments of industrial robots increased by more than 60% in 2000–2012 at 4.8% per annum. In China, a labour abundant country, robot installations grew by 25% per annum in the period 2005–2012.[6] This was apparently at least partially a reaction to the rise in real wages in China.

Various assumptions can be made about the rate at which robots will displace humans in production but if we assume the number of robots in use to be growing at a constant rate of 4.8%, the average annual growth rate for the world as a whole in 2005–2012, the effects of robotization on unemployment can be serious. The rate of growth of robots in use might even exceed 4.8% per annum: information about the superior efficiency of robots in a large variety of work environments, arising partly due to the fact that they can often be trained much more effectively and speedily than humans, might reach more firms; economies of scale, observed in industrial and sometimes agricultural production, will also apply to production of robots and bring down prices at which robots would be available to consumers and producers requiring their services; robots would become more productive and more effective substitutes for human labour over time because of innovation and learning-by-doing in the robot-producing industry.

The development of the Robot Operating System (ROS) has helped in generating various software for powering robots to do different tasks. ROS runs on principles similar to those for operating systems such as Microsoft Windows or Google's Android. Already Willow Garage's *Turtlebot2* can move around and perform tasks in an apartment on its own, build 3D pictures, and fetch various things, including food, for human beings. This robot uses ROS and was preceded by *Turtlebot*, a more primitive robot which sold for around 60% of its price. The *Turtlebot* sequence was developed by Melonee Wise and Tully Foote in November 2010.

I now look at the varied consequences of the surge in robotization in manufacturing. The first of these relates to reshoring. Automation/robotics has given U.K. and U.S. textile and apparel exports a new lease of life. Between 2009 and 2012, U.S. exports rose by 37% (Clifford, 2013) while a doubling took place in the U.K. between 2003 and 2013 (Ford, 2015).

Reshoring, though capital-intensive, creates jobs in service sectors that support the concerned manufacturing industry. Reshoring is driven by rising offshore labour costs in manufacturing hubs such as China and by the fact that location of factories closer to consumption hubs such as the United States and Europe reduces transportation cost. In April 2012, the Boston Consulting Group concluded that nearly half of companies with sales exceeding $10 billion were either in the process of or considering bringing back factories to the United States (*The Economist*, 2013). This is to be expected as the consumption per capita is very high in the United States, which offsets unfavourable incidence of factors such as the density of population in the United States being much lower than that in developing countries such as Bangladesh and India. Similarly, in 2014 a survey by U.K. manufacturers organization, *Engineering Employers' Federation* (EEF), found that one in six manufacturers was reshoring elements of its business for obvious reasons (Groom and Powley, 2014). The above discussion implies that robotization will impact employment in developing countries through two channels. First, robots can replace humans in factories located in these countries because it is easier and often less costly to manage robots than humans. Second, reshoring induced by robotization could mean that the locus of production can shift from developing to developed countries which because of their much higher consumption per capita, a result of significantly higher per capita income, are still the consumption hubs and likely to be so in the near future. No wonder, China lost about 15% of its manufacturing workforce or about 16 million jobs between 1995 and 2002 (as reported in Ford, 2015).

The mentioned trend of increasing robotization in Chinese manufacturing is being led by companies such as Foxconn, which is the biggest contract manufacturer of Apple devices. China is also heavily invested in by European robot manufacturing companies such as the ABB group and KUKA AG which build robots to feed Chinese industry. Robotization is an attractive proposition in China because of the artificially low interest rates prevailing there.

In recent times, even Chinese companies are investing abroad in countries such as the United States to set up robotized factories. An example is the development of the *Sewbot* (Bain, 2017) which can stitch garments at a rate that is 17 times that of a skilled worker. The Sewbot technology will be applied in a factory in Arkansas to provide a more efficient alternative to human sewing of shirts: 24 "Sewbots" will now produce an *Adidas* shirt every 30 seconds (The Associated Press, 2019). The total employment generated for Arkansas is estimated to be a tiny 400 persons. The personnel cost for each T-shirt will be extremely low at 33 pence per shirt, lower than that corresponding to labour-intensive manufacturing (Innovation in Textiles, 2017). It is almost certain that this negligible employment generation in Arkansas will cost Asia dear in terms of employment loss. At present, Asia accounts for a lion's share of *Adidas* factories (around 600 in all with 337 in China, 99 in India, 79 in Indonesia, and 77 in Vietnam). With automation of the kind mentioned above, prospects for those working in *Adidas* plants in developing countries do not look rosy: massive job cuts and closure of plants might be on the cards (Grahame, 2017).

It is important here to point out the obvious: the impact of large and positive growth rates of robotic employment exceeding those of human employment will gradually bridge the gap between these two types of employment. Thus, as a phenomenon, employment of robots will become more significant over time and the same can be said about the loss in human employment due to robotization.

There are many reasons why robotization is preferred to human employment in manufacturing, as has already been indicated: first, robots are easily reprogrammable and can be made to work ceaselessly, even in 24-hour shifts; second, their efficiency would eventually surpass that of humans if that is not already the case; third, their use does not involve managerial issues such as dealing with dissatisfaction among human workers, coping with unrest among workers and the resulting implications for cessation of work, and the need to look after the mental health of human labour.

To highlight the last point, consider what has been happening to Foxconn workers recently. Li Ming, aged 31 and a Foxconn worker, committed suicide by jumping from a building in Zhengzhou, China, where he had been working for Foxconn. Note that the lack of desirability in working conditions, which induced Li Ming to jump, affected a large proportion of Foxconn workers: about 350,000 Foxconn workers reportedly work in the city of Zhengzhou producing half of all iPhones, at a rate of 350 per minute (see Gonzalez, 2018 for details).

Poor working conditions for labour and worker-suicides have been an ongoing problem for Apple Inc. and Foxconn: demanding production targets and long working hours tailored to meet these targets, as well as sweat shop conditions to minimize costs led to a spate of worker-suicides and protests during 2010 and 2012. There were also protests by organizations looking at human rights issues. Apple later claimed that it had addressed problems in its supply chain which had caused these human rights issues. However, these problems later came back to haunt Apple and the workers once again: in 2017, Apple and Foxconn admitted to abusing Chinese student interns by making them work in 11-hour shifts to assemble the new iPhone X prior to its release. This was obviously in violation of local labour laws (Gonzalez, 2018).

It is obvious that Foxconn was subjecting workers to abuse to maintain its dominant position in the market; otherwise, it would not have subjected itself to the risk of attracting such adverse publicity. But with robots coming in, Apple and Foxconn obviously see this as an opportunity for maintaining their dominance without attracting the adverse publicity they have experienced while using human workers to maintain demanding targets. Therefore, employment of robots by Foxconn and other companies will probably continue to increase in an environment where only the fittest survive.

Most importantly, it seems that the cost of training robots is much less than that of training humans, as illustrated earlier by the case study of training Baxter robots to do specific jobs. As is well known, training a team of humans to each do the same job can be quite cumbersome and requires exclusive sessions customized to deal with worker-specific strengths and weaknesses: unlike robots, among humans there is a large variation in speed and ability to pick up skills and adapt to new working environments. A lot of handholding is this required.[7]

At present, there are many formal manufacturing activities which are machine-driven but not robotized. As mentioned, while fully robotized production does not require human intervention for significant lengths of time, machine-driven activities which are not fully robotized are based on the active collaboration of man and machine throughout the duration of production. The participation of humans in the case of fully robotized activities will lie largely in programming robots and in switching these "on" and "off" even though this role too would shrink for activities that have been going on for some time: as mentioned, machine learning will increasingly imply self-programming of robots, i.e., they would respond to environmental conditions or job requirements by changing the programs which form the basis for their own working.

All available robotized substitutes for human activity might not be in active use as the decision to invest in these is made on the basis of relative costs and efficiency. However, even if a robotized substitute is not in use currently, it might well come into use in the future because of Moore's law – the statistically observed annual doubling of computing power associated with a computer or robot – and the economies of scale associated with production of industrial robots.

A basic discussion of Moore's law is provided in Ford (2015) but a more detailed discussion is provided in Schaller (1997). Moore's law originated around 1970 and its simplified version says that the processing power of computers would double every two years. Moore's law works through the possibility of crowding more and more transistors on to the same integrated circuit and thus reducing the cost incurred per transistor (Rotman, 2020). In other words, the cost incurred per transistor associated with many transistors crowded on to the same integrated circuit (chip) is given by $\frac{F}{x} + y$ where F is the fixed cost associated with production of a chip of unit area, x is the number of transistors placed on this chip which has been increasing with time, and y is the cost of a transistor without accounting for the cost of the chip on which the transistor is placed. Assuming a constant value of F and the observed declining trend in y, the total cost incurred per transistor decreases as x increases with time. This implies that as time has gone by it has been possible to reduce cost and increase processing power of computers such as desktops and laptops. To see this, consider the cost of n chips of unit area, with x transistors mounted on each chip: $nF+nxy$. As x increases, total processing power nx can be driven up significantly even when n decreases significantly but at a rate which is overwhelmed by the rate of increase of x. Given that y is small and decreasing over time, the yearly increase in nxy caused by the increase in processing power nx cannot exceed a small positive amount which is easily overwhelmed by the decrease in nF caused by the mentioned decrease in n, given the large magnitude of F. Thus, we are able to get a higher processing power in computers and robots at a lower cost.

With computers able to process more information there arises the possibility of these becoming more efficient workers than humans and displacing them from the work force. Gradually, however the possibilities of cramming more and more transistors by building smaller transistors have become exhausted. The only way in which the time trajectory of processing capacity of computers can uphold Moore's law is through the writing of more efficient software. But since software caters to specific problems, the hardware industry would have to concentrate on building hardware that caters to specific software. Thus, Moore's law would hold only for certain types of computing problems but not for the entire mass of computing problems.

Thus, a decline in the cost of production of robots and an increase in the efficiency of robots may drive down the cost of undertaking robotized manufacturing activities, resulting in the overall expansion of robotized activity in the economy and a tendency for human–driven activity to contract. It must be remembered though that the overall global scale of economic activity is expanding all the time, with large economies such as India and China clocking growth rates of gross domestic product over 6%. The human–driven activities which support robotized economic activity might have to expand if robotized activity expands. Thus, a safe prediction would consist of forecasting an increasing share of global GDP produced through robotic activity, not an absolute contraction of human-powered production activity.

On the other hand, informal manufacturing might continue as it is based on a low level of mechanization and high labour intensity: it is often carried out by the economically underprivileged who do not have the access to credit and finance to buy expensive machines or robots and are content with a low mixed income[8] for their labour. However, as the variable cost of production in the robotized sector declines due to advances in robotization, the products of this sector might in course of time become cheaper than even the substitutes produced by the informal sector so that even the poor might start buying from the former sector. Thus, the scope of the informal sector might get purely restricted to services such as prostitution, low-cost massage services, caregiving and domestic help, and the like. However, with robots enabling employed humans to follow a lighter work schedule, the scope for sale of these services might however increase as time available for consumption increases.

The impact of robotization on the services sector

In regard to the formal service sector, the scope for robotization is immense though the exploitation of this potential has only started in recent times. However, such robotization is already significant.

In the hospitality, old age care as well as hospital sectors, humans are being substituted efficiently or supplemented by robots. In the retail industry, self-check-out lanes have made major inroads into employment of humans. This includes replacement of floor assistants by self-vending kiosks. Even jobs that conventionally require a lot of human capital are being substituted by robots. For example, software-driven "artificial" sports journalists are reporting on sports proceedings without any interference by humans at all. Clusters of computers are belting out novel and original strains of classical western music. Software-driven artists are depicting human emotions on virtual canvases.

With robots being able to perform mechanical as well as many creative tasks, the prospects for human labour might look grim. However, there are still many jobs, such as the top management positions in businesses and those in the hospitality sector involving complex communication with clients, which are not yet under threat of being captured by robots. Note that with robots performing many activities that humans perform at present, even employed humans will have more time for consumption and leisure. Thus, businesses providing entertainment/hospitality such as restaurants, hotels, and those selling cultural products are expected to bloom. Even if robots are used for activities within the hospitality sector, many of the services in that sector which require interpersonal skills (emotional intelligence and ability to answer complex queries of clients) will still require employment of humans. With caregiving and hospitality sectors expanding, partial robotization will still be consistent with increase in human employment.

Let us look at the expanding role of robots within the hospitality sector. In hotels, robots can be used to provide information that can be of help to guests during their visit (see Revfine, 2019 and Socialtables, 2019, for details). More formally, robots can be of help in providing front desk services in place of receptionists, services in transporting and storing luggage that make use of technologies such as collision detection and Wi-Fi to avoid obstacles, as well as check-in and check-out services which include checking of identity documents and provision of essential information such as procedures for entering hotel rooms and exiting hotels after stay. The development of speech recognition abilities has made it possible for robots to interact with guests and answer their questions. Machine learning makes it possible for the robot to learn from each interaction and improve upon its answers. An advantage that robots have over human assistants is that they can provide information services in different languages. A related development is in travel agencies. These travel agencies are making use of "chatbots" to collect necessary information from customers and make their hotel and flight bookings.

It should be noted that all hotels might not undertake such robotization as many customers might prefer interacting with humans: the warmth of a human smile, empathy, ability to answer complex questions, etc. have evidently not been picked up by robots, an inability which leads many hoteliers to completely bank on human–human interactions in the process of providing hospitality.

More impressive are the ways in which robots are being used in restaurants to cut prices as well as displace human employees. Consider dining services provided through restaurants and fast food kiosks exemplified best by the *Kura* "revolving sushi" restaurant chain which recorded a boom in a stagnating Japanese economy in the first decade of the 21st century. The discussion in this book is based on a detailed investigation of the rise of *Kura* published in *The New York Times* in 2010 (Tabuchi, 2010). In order to minimize costs, this restaurant chain has invested in sushi-making robots which have replaced the traditional sushi chefs. There are very few waiters; the flocks of waiters which should have characterized the sale of a huge volume of sushi in a sit-down restaurant have been replaced by conveyor belts. Diners can pick up plates laden with sushi from conveyor belts whose motion in *Kura*'s 250 odd restaurants is monitored and controlled by remote managers stationed at three control centres situated across Japan. The motion of these plates is managed after food prepared by robots is put on the belt so that diners can get the food requested by them in quick time (discussion based on Tabuchi, 2010).

A plate of sushi cost a little more than a dollar in 2010 in a *Kura* restaurant which was much lower than the exorbitant prices characterizing specialized sushi restaurants. This is the reason why *Kura*, after its inauguration in 1995, boomed in the first decade of the 21st century in a shrinking restaurant sector in a stagnating economy where the common man was spending less on eating out, given weak wages and a fall in the annual private sector pay by 12%. It swam against the dining-out slump on the basis of its low prices and a single-minded pursuit of efficiency by its managers. For example, according to the Foodservice Industry Research Institute located in Tokyo, in 2009, restaurant revenue fell 2.3% to 23.9 trillion yen which was less than the peak revenue of 1997 by 20%. While conventional restaurants were closing down *Kura* was booming.

Some more elaboration regarding the dining services being offered at *Kura* is in order. *Kura* has done away with management staff devoted to each restaurant. Rather there are central control centres which have video links to various restaurants. The provision of service at various restaurants in each region is managed by a small group of managers at the central control centre dedicated to that region. This group monitors everything: from the size and apparent quality of servings to the décor in the restaurants that belong to the region administered by them. Within each *Kura* store, services are managed through automation: touch panels are given to diners to order dishes which reach the diners at their tables through the mentioned conveyor belts linking them to robots working in the kitchen. The plates carrying food are then counted to calculate the bill, cleaned using fluids and returned to the kitchen using a mechanized process. The substitution of human staff by machine is high: each restaurant capable of seating 196 people is managed by six human servers and a minimal kitchen staff.[9]

Robotized services that prepare burgers from scratch and other fast foods are also coming into popular use to displace human workers in order to reduce operational costs. A good account is provided by a 2018 newspaper article in

The Sun (Keach, 2018): in early 2018, *Flippy* the burger-flipping robot took up residence in a Californian restaurant; the burger chain Wendy's begun installing self-cleaning ovens in some stores while the restaurant chain Arby's started using smart ovens to cook roast beef overnight instead of having human employees do the same by arriving early in the morning. A case of almost complete robotization is that of *Momentum Machines* which are using machines to prepare burgers without any human intervention. There is a belief in the company that elimination of labour cost will allow it to add quality ingredients and thus facilitate the provision of gourmet burgers at fast food prices (Ford, 2015).

In the retail sector, automation is again resulting in a displacement of human employees. According to experts, automation is gathering momentum in the retail sector and could be the cause of an increase in joblessness or a stumbling block to efforts to bring down the same. This industry has been known to employ about one in ten Americans but this state of affairs might not continue for long as employers try to sell more products using fewer employees. Substitution of humans with more efficient robots is taking place even at the shipping and warehousing stage. On the retail floors human customer service representatives are increasingly giving way to virtual assistants, and check-out clerks to vending kiosks and self-server machines. Products which can be bought in this manner include iPods, bathing suits, sunglasses and razors, and even prescription drugs. In the future, even automatic purchase of motorcycles and cars might be possible (Ford, 2015).

Semuels (2011) further goes on to point to developments on the anvil: gas stations which dispense gasoline, snacks, and bottled foods without any human clerical assistance through the swipe of a credit card. These are designed by the Corona vending machine firm *AVT Inc.* which is sure that such stations are a good answer to the woes of handling human capital: the vending machines would work 24 hours a day and would give instant feedback regarding inventory and sales. Important changes might soon be afoot in the nature of vending machines: they may be able to provide health and nutrition information, serve cooked food and be connected online (Lee, 2016). Thus, the popularity of vending machines might increase further.

Semuels (2011) points out to a discernible slump in employment in 2011 in the United States from the January 2008 figure by 7.5%. Meanwhile the amount transacted through vending machines registered a rise: an increase of 9% from 2009 to reach $740 billion in 2010. Experts projected a further rise to $1.1 trillion by 2014. Though some of the decrease in employment was due to the great recession, it is usually seen that a post recessionary period associated with improvements in labour-saving technology is hardly associated with displaced workers coming back to work.

In some supermarkets, the self-service lanes that have been installed are being thronged by customers and do not involve the labour costs of employing human cashiers. The American subsidiary of British retail giant Tesco has converted all check-out stations at its outlets into self-service kiosks which have reportedly impressed customers with their speed and shorter lines. These also protect the privacy of the customer, especially while buying personal items such as contraceptives and underwear.

Employment in conventional movie and book chains and other similar outlets has come under the negative influence of the spread of online websites – for instance, movie and book chains, *Blockbuster* and *Borders*, have been forced to shut their business by the increasing popularity of online sales. Such online sales not only imply job losses in retail sales but also in packaging of orders as packaging linked to online sales can be done in warehouses which are much more amenable to robotization than shop floors.

Consider the recent case of supervision of 189 *Redbox* movie rental kiosks in the Chicago area. These kiosks were managed through remote maintenance with a staff of seven. Compare this with that of the case of *Blockbuster* where 60,000 employees were employed in 9,000 stores in the United States (as reported in Ford, 2015). This translates to 6.67 employees per store as opposed to less than 0.04 employees per kiosk for *Redbox*. In other words, the labour intensity at *Blockbuster* was approximately 150 times that of *Redbox*, assuming that one kiosk can provide the same service as a store.

There is further scope for robotization in the sales sector in the future. As robotic technology improves so will robots' powers of mobility and other types of useful physical dexterity as well as visual recognition, all required for robots to perform more functions in shops such as packing, stocking, and collecting items from shelves for customers (based on discussion in Ford, 2015).

The scale of mobile ordering has been registering a steep increase over time: for example, restaurant mobile visits in the United States – use of a mobile app to pay in advance for food which can be picked up as soon as the customer reaches the restaurant – increased by 50% in 2017. Mobile ordering of food for home delivery has also become very popular in the urban centres of various countries including developing countries such as India. This involves using an app which connects the person placing the order to a food deliverer that can fetch food for her from a range of restaurants. A report by Google and Boston Consulting Group (BCG) projects that India's online food industry will reach a scale of $8 billion in 2022, growing at a rate of 22% per annum (Mint, 2020). This industry is present in more than 500 cities in India. The rise of this industry in India should be seen as associated with the global growth of this industry: revenue generated globally was $82 billion in 2018 and given an annual growth rate of 14%, is expected to double by 2025. China accounts for more than 40% of global sales of this industry (Singh, 2019).

A major boost to the online food industry is the emergence of cloud kitchens, also mentioned by Singh (2019). Each cloud kitchen prepares a variety of foods for different brands – virtual restaurants which are not physically located. The cloud kitchen's premises are structured so as to facilitate easy access by drivers engaged in food delivery.

Convenience stores are also registering a precipitous increase in mobile ordering over time. A sum of $813 billion was spent during the calendar year 2020 alone on online shopping in the Unites States, a growth of 42% from 2019, in part attributable to the Covid pandemic (Perez, 2021) In India, with four times the population of the Unites States but a much lower per capita income, overall online spending is expected to reach over $130 billion in 2025 (Mint, 2020).

Next consider caregiving services provided by robots to the elderly. We are now seeing the beginnings of caregiving by robots through the catalysis of social interaction which has already started to supplement caregiving by humans of an ageing population (Petrecca, 2018; Dormehl, 2019). A significant variety of these robots has already been pressed into service: AvatarMind's *iPal* will eventually monitor for falls; Catalia Health's *Mabu* can ask questions like a nurse to monitor the well-being of elders. *Paro*, an interactive baby seal which can be purchased for $6,000 is another robot that provides emotional support. It is the size of a large stuffed animal and responds to touch, noise, light, and temperature by moving its body parts and making sounds. By helping to improve the mood of its users, this robot offers some relief from anxiety and depression, a frequent malaise among elders in Japan, a rapidly greying nation.

Research has suggested that interaction with *Paro*, far from replacing human interaction, can increase socialization among the elderly in locations such as care homes. The mechanism is simple: the mood of being depressed and withdrawn is improved and therefore the chances of humans, especially the elderly, of interacting among themselves are improved.

Companies are still looking to make improvements in their elderly focused software. The *iPal*, for example, may soon be able to remind the elderly to take medicine and provide them with content for entertainment, puzzles, etc., to keep them engrossed, and a whole lot more.

In addition to the robots named above are those which actively monitor pathological conditions: *Mabu* (above), a 15-inch yellow robot helps patients with congestive heart failure by keeping track of medications and activity level. It can also screen for anxiety and depression.

These robots cannot help the elderly to carry out physical tasks like preparing meals. The elderly would still largely need humans to carry out tasks, though the physical disabilities of the elderly themselves can be alleviated or rendered less consequential by bringing robots into use: tools like robotic exoskeletons can help people with limited mobility to manoeuvre themselves more easily; voice recognition technology can be used in homes housing the elderly to operate lights, televisions, and other appliances. Additionally, while robots can make a contribution to elderly care through social assistance,[10] at the same time it is being felt by the medical community that algorithm-driven robots are not a complete substitute for the empathy and kindness provided by humans. Thus, human caregivers will continue to be required for the elderly for both human and social assistance as the number of the elderly registers a very significant growth in the coming years. Petrecca (2018) points out that according to the United Nations, the number of people 60 and older in the world is expected to reach 2.1 billion in 2050 from its level of 962 million in 2017. Thus, in spite of advances in artificial intelligence and robotization, labour demand for caregiving of the elderly will increase significantly in the next few decades. This will be a major source of employment and help to neutralize job displacement in other sectors.

From a field where human employment is not under great threat, we now move to a field where human employment is already being reduced significantly due to the impact of cyber technology. Like caregiving to the elderly, hospitals too involve the provision of labour-intensive services to people such as those provided by orderlies, nurses, and doctors. An advantage of robots displacing humans in hospitals is that robots can perform tasks, ranging from routine menial tasks to diagnosis to actual complex surgery, with far greater precision.[11]

Consider the case of *Noah*, a robot that has started doing the rounds in the Guangzhou Women and Children Medical Center and has been fetching medicines, carrying documents, and other things from point to point. It is far more efficient than a human as it can carry a weight of 330 kg around the hospital, far more than a human nurse or orderly can carry. By concentrating on routine tasks *Noah* provides room for nurses to carry out more important tasks. *Noah* and other robots have a GPS system which helps them move around the hospital while not bumping into human beings and other obstacles. *Tug*, a more advanced robot uses high tech lasers to map out space in front of it and thus avoids bumping into objects.

Surgical robots are also being employed by surgeons to undertake precise surgical manoeuvres such as cuts and incisions. Usually these are not "robots" in the real sense of the word as they cannot function without human guidance, though there are some surgical robots which can perform entire surgical procedures on their own. One factor limiting the spread of these robots to various hospitals is that these are very expensive, with cost running presently into millions of dollars. However, as the technology underlying these robots develops, these will surely become less costly with use becoming more common.

Research into how humans view automated healthcare shows that people are slightly sceptical of availing of healthcare from an automated system, but this scepticism does melt away with time, especially when the use of an automated system helps save on expenses. A factor which influences the acceptance of surgical procedures by robots is the invasiveness of the surgery. The greater the invasiveness, the more difficult it is for humans to accept robotized versions of surgical procedures. For example, a survey shows that two-thirds of respondents did not want a robot to handle an invasive procedure like a root canal. When it came to robotic cleaning and whitening, only 32% said they would decline.

Our reviews show that employment of humans is being impacted by artificial intelligence through the length and breadth of the service sector. In many cases, such as hospitals and hotels, robots are acting as substitutes for human employees, though in others such as caregiving of the elderly these are only supplementing the roles played by human employees and in many cases helping them to perform their roles more effectively. Note that displacement and supplementation is not restricted to lower-level workers only. The middle-level workforce in the service and even the manufacturing sector is also being squeezed due to robotization: computerization is wiping out the roles being played by middle-level managers and supervisors. This is leading to the phenomenon of the "missing middle".[12] Furthermore, computers/robots capable of "machine learning" can allocate work to fairly highly paid executives/consultants, observe the action sequences leading to the completion of assignments, and then replicate the process, finally leading to the ouster of highly paid white-collar workers (Ford, 2015).

Thus, even some top-level job opportunities are being lost. Some of the most sought-after jobs are in the banking sector. For example, substitutable starting analysts make $91,000 in base pay, while managing directors can earn almost $1 million after yearly bonuses (Akhtar, 2019). Banks have already started making investments in automation and have started the process of using AI to mimic bank employees and automate processes, a trend which will end up displacing humans. Reports indicate that a million jobs might disappear from the banking sector by 2030.

Cloud robotics constitutes an important development with significant implications for production and distribution as well as employment of humans. This is a way of storing various software in a centralized place. Individual software can be uploaded into a large number of inexpensive low memory robots to get a task done. Once the task is done the software can be deleted and another can be uploaded to get a different task done by the same army of robots. Thus, the same inexpensive robots can perform a variety of tasks, making the possible magnitude of substitution of humans by robots far larger than before.

A further development relates to the labour intensity of human capital formation. Mass online courses offered by reputed academics are fast gaining in popularity.[13] These high-quality courses allow students in distant countries such as those in Africa and Asia to access the wisdom and knowledge of the world's foremost academics located in the better-paying West. Grading of a large number of answer scripts, an essential part of the implementation of these courses, can be accomplished by machine learning replicating the manner in which an instructor grades answer scripts. Thus, a robot can grade thousands of answer scripts in a way that mimics the grading of answers by the human instructor without the random variations characterizing human grading.

However, there are problems: the technology of monitoring students when they are taking exams is far from perfect and leaves scope for unfair means such as the use of dummy examinees. Solutions include getting students in far-flung countries to come to major cities in these countries with photo identity cards to take exams under the close scrutiny of watchful proctors. However, this solution has not been fully implemented. Thus, wariness about providing degrees to online students remains. Another problem with online courses is the difficulty in replicating the classroom interaction of universities characterized by the physical presence of instructors and students in classrooms. Furthermore, Ford (2015) points to a major reason behind the reluctance of private universities to subscribe to this model: mass online courses providing degrees leading to a dwindling of the demand for instructors and hurting the business of private universities. It can be expected though that the obstacles to mass online courses providing degrees might get surmounted in the intermediate run, say 10–15 years from now. After all, mass university education might be a good revenue tool for a reputed university catering to a very large area through online means at a reasonably low cost. However, the lesser universities might be rendered unprofitable as each of the highly reputed universities would capture a larger student base, eroding the student bases of the former type of universities. For the same reason, jobs in the education sector would be hurt.

Note that robotic activity is increasing in the following sectors, though humans still have a massive lead over robots which seems unlikely to be erased in the near future: translation,[14] writing of novels (for details, see Schaub, 2016), news reporting,[15] art,[16] and composition of music.[17] It will, however, take a very long time for robots/humanoids to match the versatility, charisma, and skill of top-level singers, novelists, poets, film stars, theatre actors, and other purveyors of the arts.

Conclusion

There is no doubt that robots will displace a significant number of humans from their jobs in agriculture, marketing, and services, given the review in this chapter of the uses these are being put to. At the same time the higher volumes of outputs associated with robotization are going to generate an enhanced demand for labour in marketing and transportation. AI-assisted workers also should see a shortening of working hours and enjoy an increase in leisure time, which will in turn lead to an increase in the demand for human capital and soft skill-intensive services in the hospitality industry and the entertainment sector.

There would be a major increase in demand for human capital in geriatric care due to an increase in the proportion of population accounted for by the geriatric age group brought about by a rise in life expectancy. With robots promising to make the elderly more social, active, and upbeat, the elderly would become ready to lead a varied life. In translating this willingness to an actual increase in activity, the assistance of young people would be required. In other words, this sector should at least in the short run see an increase in the demand for human labour which has an advantage over robots in a home environment in regard to lifting, fetching, cooking, and verbal communication, especially that conveying positive emotions and complex ideas.

Thus, we can see that though robots might be seen as labour-saving devices, their employment generates a lot of tendencies that also would result in job creation.

Appendix: the impact of robotization on the demand for labour inside an agricultural economy

Let us consider a stylized model: there are n farms in the economy, each of size 1 hectare. I assume that there is no market for the sale and purchase of land. This is not far from the truth in regard to many developing countries which are marked by an absence of formal property rights. I also assume that there is no rental market for land. Our assumptions imply that the economic cost or opportunity cost of agricultural land, defined as the value derived from the next best use of land (which the farmer has to forego when cultivating land), is zero.

Land is combined with human labour to produce an agricultural good which is the numeraire, i.e., it has price equal to unity. In other words, I measure all other prices in terms of units of the agricultural good. Thus, a variable such as wage (w) actually corresponds to w units of the agricultural good, i.e., unit labour is compensated by giving it w units of grain.

The labour employed by farmers to cultivate land is drawn out of a common pool shared with the urban informal sector where employment is provided at a low but constant wage, the marginal product of labour. As agricultural wage goes up the labour drawn from this pool by the agricultural sector for employment increases, i.e., the supply curve for labour facing the agricultural sector is upward sloping.

The urban formal sector is not explicitly modelled but we can consider it to be a price taker in the international economy, drawing qualified labour from a separate pool of workers. Note also that the amount supplied by the agricultural sector is not constrained by local demand as the agricultural economy caters to both local and international demand.

Before robotization, yield is given by A min $(1, L)$ or product of A and the minimum of the numbers 1 and L, the amount of labour employed. Thus, the amount of production resulting from employment equal to L is given by AL as long as L is not greater than 1. This is valued at AL given that the agricultural good is the numeraire. The amount spent on employment of labour is given by wL. For L not greater than 1, profit is given by $(A - w)L$ and is maximized at $L = 1$ if $A > w$ and $L = 0$ if $A \leq w$. Note that any excess of L over 1 adds nothing to production of the agricultural good and enhances the wage bill. Thus, the farmer does not consider values of $L > 1$ while maximizing profits.

To summarize, the amount of labour demanded per farm is 1 if $A > w$ and 0 otherwise. The labour demand curve gives the sum of labour demanded by the mentioned n farms at different levels of w. Thus, in the diagram above, the labour demand curve is a vertical line passing through $L = n$ for $A > w$ and $L = 0$ for $A \leq w$. Provided the upward sloping labour supply curve is located far enough to the right (i.e., the quantity supplied at various wages is high enough) the equilibrium wage, w^* will be less than A. I assume that this is the case and therefore the amount of labour employed in equilibrium is n.

Let us assume that the robotic technology becomes available at this time. For each farm, robotization or R takes on a value in the range 0–1 with 0 corresponding to no robotization and 1 corresponding to full robotization. A value of R equal to a fraction corresponds to partial robotization; a higher value of R corresponds to a greater extent of robotization. Under partial robotization some labour is obviously required for the operations that are not robotized. The overall technology available to the farmer for cultivating land is assumed to be given by $Y(R)\min(1, L+R)$ with $Y(R) > A$ for $R > 0$ and $Y(0) = A$, the amount of output that can be produced if the technology predating robotization is used after choosing $L = 1$. Note that $Y(R)$ is assumed to be increasing in R.

Note that the farmer will never employ a level of $L + R > 1$ as the excess of $L + R$ over 1 makes no contribution to output but is costly. As long as $A > w$, $L + R < 1$ can never correspond to an optimal choice of labour employed and robotization: it is always possible to increase labour employed starting from a position where $L + R < 1$ and enhance profits, given $Y(R) > w$, which in this case is always true. Hence, $L + R = 1$ or $L = 1 - R$ is a necessary condition for profit maximization.

Let c be the unit price of robotic equipment for a period, i.e., I assume that the services of the robotic equipment can only be bought for its entire lifespan which covers multiple periods and through an obligation which effectively pays c per period. In short, I assume that services of a robot cannot be hired for any length of time less than its entire lifespan. The farmer will maximize profits by choosing R from the range 0–1: she will maximize $Y(R) - cR - w(1 - R) = Y(R) - (c - w)R - w$. Here $Y(0) - w = A - w$ is the profit corresponding to zero robotization. When robotization is undertaken, profits change through two channels: an increase in R enhancing yield as $Y(R)$ is increasing in R and a change in costs given by $(c - w)R$. If $(c - w) < 0$, then more robotization adds to yield as well as reduces cost; the profit maximizing choice of R is 1. If $(c - w) > 0$ this is not the case. Let us assume that the marginal product from robotization, $Y'(R)$ is not only positive but diminishing in R. It is obviously profitable to keep on increasing R, starting from the value 0 as long as $Y'(R) > c - w$, i.e., yield and therefore revenue changes at a rate faster than cost with robotization. Equilibrium is attained at a value of R which is a fraction if $Y'(R) = c - w$. If $Y'(0) \leq c - w$ then the optimal value of R is zero and thus $L = 1$. Full robotization is optimal if $Y'(1) \geq c - w$, i.e., the rate of change of yield with respect to robotization exceeds the rate of change of cost at all possible values of R.

Thus, full robotization results when robots are cheaper than human labour or when the marginal gain from robotization in terms of yield is always greater than the marginal cost incurred; zero robotization occurs because of gains from robotization in terms of yield not being large enough in comparison to costs incurred at the margin at all levels of robotization; and partial robotization occurs when the mentioned gains from robotization overwhelm the mentioned costs to start with but are not strong enough to exceed these costs at all levels of robotization.

Given that the wage just before the introduction of robotization is w^*, the amount of robotization undertaken will be given by a fraction R^* satisfying $(c - w^*) = Y'(R)$ if such a fraction exists. This will be the case if $Y'(1) < (c - w^*) < Y'(0)$, the other cases being full robotization $(Y'(1) \geq (c - w^*))$ or no robotization $(Y'(0) \leq (c - w^*))$. Assuming that the first case holds, the aggregate amount of labour demanded is given by $n(1 - R^*)$ which is less than n, the aggregate labour demand predating robotization. Note that n is also the labour supplied at w^* before the robotic technology is made available which implies that an excess supply of labour is generated due to robotization that brings down the wage rate. However, note that the demand for labour is perfectly inelastic for wages below w^*: once the promise to pay c per unit of robotic equipment per period for the entire lifespan of the equipment is undertaken through purchase of the robot it cannot be broken; hence, any decrease in robotization by ΔR on a farm in response to a decrease in w will not result in any saving but an additional expenditure of $w\Delta R$ on labour that replaces robots as also a decrease in yield. In other words, the optimal change, caused by the decline in wage rate, in robotization from R^*, $(\Delta R)^*$ is 0.

Thus, the demand curve for labour is perfectly inelastic for wages below w^* at $n(1 - R^*)$ and downward sloping for wages above w^* (see Figure 3.1). The mentioned excess supply of labour at w^* therefore drives the equilibrium wage to below w^* with labour demand fixed at $n(1 - R^*)$. The new equilibrium is given by the wage rate, $w^{OR}(< w^*)$, and employment of labour, $n(1 - R^*)$ (see Diagram 3.1), given by the intersection of the labour demand curve (in green), $D^{OR}(w)$, and the labour supply curve, $S(L)$, with the superscript OR referring to the scenario in which robots can only be owned but not hired.

We can consider an alternative scenario: the farmer has the option of hiring the services of robots in every period by just committing to the per period costs, which we can again refer to as c. In this case the introduction of robots again drives labour demand to $n(1 - R^*)$. Subsequently, excess supply leads to wage rate again falling below w^*. In this case, a decline in robotization starting from R^* leads to cost saving of $c - w$ exceeding revenue loss of $Y'(R^*)$, i.e., a decline in robotization and an increase in labour employed is profitable. Denoting R^{HR} as the equilibrium number of robots reached through a fall in the equilibrium

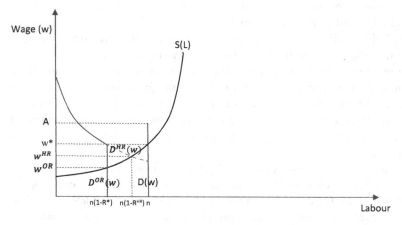

FIGURE 3.1 Comparison of employment and wage rates in pre- and post-robotization scenarios.

wage below w^* to w^{HR} the equilibrium amount of labour, $n(1 - R^{HR})$ lies between n and $n(1 - R^*)$ and the equilibrium wage rate w^{HR} satisfies the following condition: $w^{OR} < w^{HR} < w^*$. This can be observed in the diagram from the intersection between the curve consisting of grey and dotted segments, $D^{HR}(w)$ and the labour supply curve, $S(L)$.

Thus, robotization is always associated with a decline in wage rate as well as employment, but the magnitude of this decline is always greater in a scenario in which a robot has to be bought as compared to a scenario in which a robot can be hired at a per period price equal to that corresponding to ownership.

Notes

1 The median income is the income level associated with the 50th percentile of the income distribution, i.e., if you arranged individuals in a population in ascending order of their incomes, the median income would be the income level dividing this ordered population into two equal parts. Given that the income levels in the upper 50% of this income distribution are higher than in the bottom 50%, the per capita income would be higher than the median income. As income inequality rises, those in the top 50% account for a growing share of national income which pushes up the gap between per capita income and the median income. This is exactly what happened in the United States in 1973–2003.

2 The discussion in this paragraph and the subsequent one is based on Ford (2015).

3 Details can be found in the Vision Robotics Corporate website and Singularity Hub (2012).

4 This process is described well by Ford (2015): a computer model of a tree is constructed by the robot using machine vision and indicates the location of each fruit. When this information is passed onto robotic arms it uses it for rapid harvesting. The whole process prior to harvesting may be described in more detail as follows, using the discussion in Marr (2019): a sensor detects the product (say the fruit) and triggers a light source to illuminate it and a camera to capture its image. A device called the frame-grabber then converts the image into digital information. A software then compares this digital information to a set of predetermined criteria which helps ascertain whether the fruit or flower is ripe for harvesting and whether the robotic arm should swing into action. This process is uncannily similar to how humans pick fruits or flowers: the eyes detect a fruit or flower, inspect it more completely to judge its characteristics, which are then compared to a set of standards stored in their brains to determine whether harvesting is justified.

5 This would always be the case if robotic labour is measured in terms of the human labour it displaces.

6 The information in this paragraph has been obtained from Ford (2015).

7 The discussion in this paragraph uses information provided by Ford (2015).

8 It is the combination of wage, rental, and profit incomes earned by those making their living in the informal sector.

9 The discussion in this paragraph is based on Ford (2015) and Tabuchi (2010).

10 Social assistance here refers to exhibition of social behaviours by robots such as recognizing, following, and assisting their owners and if possible, engaging in conversation (KPMG, 2016).

11 A discussion of how robots have been rendering services in hospitals is provided in YellRobot.com(2018) and Milner and Rice (2019) and has been utilized in this book.

12 Computers and the use of the internet has enabled high level white-collar workers to perform the chores of communication, computation, and accounting themselves.

This phenomenon has been explained at length first by Autor (2010) and then by Jaimovich and Siu (2012) and implies that top management does not need the assistance of middle-level managers and administrative staff. Such middle-level jobs have therefore in many cases ceased to exist.

13 The story of the increasing popularity of these online courses is narrated by Ford (2015). Salmon (2012), Selingo (2013), and Chafkin (2013) discuss the success of the first major mass open online course offered by two computer scientists, Sebastian Thrun and Peter Norvig, at Stanford University which was completed successfully by 23,000 people drawn from all over the globe who received certificates from Stanford University.

14 The increasing effectiveness of translation by AI is discussed in Hicks (2018). While corporations such as Microsoft have announced the development of AI software achieving human effectiveness in translation, the newly developed capabilities will have to be put to varied use by others before these claims can be fully accepted.

15 Software such as StatsMonkey (for sports journalism) and the more widely applicable *Quill* and *Heliograf* can construct news stories through the extensive and fast scouring of databases. For more discussion see Carr(2009), Ford (2015), and Moses (2017). Whether software can finally write articles with high analytical content will only be revealed by future developments.

16 For example, a software called *The Painting Fool* (see details of the website in references), as discussed by Ford (2015) and Shubber (2013), can identify emotions in photographs and use these as input into abstract paintings.

17 Artificial Intelligence can now compose music. For example, *Iamus*, an algorithmically powered cluster of computers, has composed a symphony which has been performed by the *London Symphony Orchestra* and received enthusiastically by some reviewers. For details see Smith (2013) and Ford (2015).

4

MACHINE LEARNING AND THE ECONOMY

Introduction to machine learning and its types

Machine learning is one way to generate artificial intelligence which facilitates the completion of tasks by computers that have until now been carried out by humans. These tasks span a very wide range from those that help to manoeuvre driver-less cars through everyday traffic consisting of animate and inanimate objects without collisions, to translating speech and making predictions regarding phenomena on the basis of available data. Our discussion of the technical aspects of machine learning is based on that given in Heath (2018).

Machine learning is not the only way to generate artificial intelligence in computers. Rather it may be defined as the generation of artificial intelligence by feeding data into computers. While traditional computer software consists of rules provided by humans to computers to accomplish a particular task, machine learning only provides data to a computer system on the basis of which rules for accomplishing the assigned task can be generated by the computer system on its own by invoking a few basic algorithms. Algorithms, in regard to machine learning, are a few simple instructions which enable computer systems to learn from data and then accomplish various tasks, even fairly difficult ones (Deangelis, 2014). In this manner, the computer system can complete a variety of tasks for which it has not been programmed.

Machine learning is of various types: supervised learning, unsupervised learning, semi-supervised learning, and reinforcement learning. Supervised learning involves the feeding of a huge amount of labelled data into a computer (training). For example, the computer system might be exposed to a huge amount of data consisting of labelled images of apples and guavas. The computer system uses an algorithm and labelled data to learn how data regarding the image maps into a "label", i.e., the name of the fruit. It thus teaches itself to recognize

DOI: 10.4324/9780429340611-4

how the label varies with the bundle of characteristics of the image of the fruit (colour, shape, etc.). The computer system thus acquires the ability to match the bundle of characteristics associated with the image of any apple or guava to the correct label describing that fruit.

After the computer is fed with labelled data the computer system is presented with unlabelled data (images in this case) and asked to generate labels. The performance of the computer system is then evaluated by generating man-made labels for the unlabelled data, matching the man-made labels with the labels generated by the computer system and thus determining the proportion of cases in which the man-made labels and the computer-generated labels match. This proportion is a measure of accuracy and if this is unsatisfactory, methods to generate more satisfactory results include (a) tweaking of the algorithms to generate labels from the data regarding images and (b) increasing the amount of labelled data used for training. The evaluation stage ends when the computer system is able to generate labels for given unlabelled data with high accuracy. It thus reaches a stage whereby the labels that it provides to data (images) can be treated as "true labels" which are generally not contested and can be used for further use and processing. In our example, where unlabelled images of guavas and apples are given, it is then able to label images with an acceptable degree of accuracy.

A larger project would involve imbuing the computer system with the ability to correctly label each type of common fruit through machine learning. Thus, through machine learning the computer would be able to generate labelled picture books of fruits. Moreover, after the ability to label is picked up by computers, they would be able to generate these picture books much more quickly and at a far lower cost than humans. As a result, the number of these picture books available for human consumption would increase.

It has to be noted though that the amount of labelled data that needs to be fed into the computer system is often quite huge. These labels need to be manually generated and their creation thus necessitates a huge amount of labour. After the labelled data set, for feeding into the computer system, is created and the training of the computer system on the basis of this labelled data completed, the future demand for labour is reduced as computers start acting as substitutes for human labour labelling data.

Note that labelling is a type of prediction – it predicts on the basis of an image the name humans attach to the image – but there can be many types of predictions that supervised machine learning can generate. Consider, for example, the amount of soft drink consumed in a city at a given temperature. The computer system is trained on a data set which gives amounts of soft drinks consumed at various temperatures. The computer system can then, for example, try to fit a linear relationship to the scatter plot which captures the data set and determine the error associated with each data point by noting the difference between soft drink consumption predicted by the linear relationship at the given temperature and the consumption level actually recorded. An algorithm can be used to add functions of the absolute values of these errors to generate a "sum" – for example, squares of errors can be added or just the absolute values can be added. The computer system experiments with the slope and position of the line representing the relationship between temperature and soft drink consumption until the mentioned "sum" is minimized. This relationship is used to make predictions. Note that under machine learning, the computer system is equipped with tools such as algorithms to make predictions but to a large extent it exercises some autonomy in making these predictions.

In the case of unsupervised learning, the computer is given data so that on the basis of algorithms it can detect patterns in the data for categorization on the basis of similarities: for example, flat addresses can be bunched according to the rent category these belong to or other similarities in the data pertaining to flats; or stories posted on a website can be bunched by topic. Again, notice that such learning enables a lot of time saving by humans; in regard to the example elaborated above, the consumer interested in renting a flat from a given rent category in a given geographical area in a city would save a lot of time on account of unsupervised machine learning because the computer system itself would be able to present the flat addresses available for rent, from a narrow geographical area, which belong to the rent category in which the consumer is interested. The time-saving itself would be transformed into economic benefits as the consumers would be able to use the saved time for productive economic activity. A similar benefit would be generated in the example of "stories bunched by topic" as the consumer would be able to direct her attention to the stories pertaining to the topic that interests her without incurring any time costs in identifying and compiling stories.

In the case of semi-supervised learning, the computer system is trained on a data set consisting of labelled as well as unlabelled data. At first the computer is exposed to labelled data, such as labelled images of fruits, so that it can generate labels for unlabelled images. These are referred to as pseudo-labels. Images carrying labels and those carrying pseudo-labels then constitute a larger data set for training the computer system to perform the task of labelling accurately.

Note that semi-supervised learning is cheaper than supervised learning as the amount of manually labelled data used for the former is smaller. On the other hand, because manually labelled data is more accurate than pseudo-labelled data, supervised learning is qualitatively superior to semi-supervised learning.

In the case of reinforcement learning, any task is viewed by the computer system as a game. These tasks include actual video games. The computer system views the game in terms of its existing visual state, the actions undertaken by the intelligent entity (computer system or human) and the scores registered as a result of these actions. Viewing more and more cycles of the game helps the computer system accumulate data on the basis of which it can generate a model for maximizing the score, given the state of the game. The model can then be implemented by the computer or a computer-driven robot to generate high scores in the game.

Reinforcement learning is built on the principles governing human and animal behaviour. If we observe our own actions and the actions of humans around us, it is easy to infer that the nature and frequency of such actions is governed by the magnitudes of rewards and punishments. Thus, a human can be modelled in simple turns as an automaton which tries to get the best outcomes through its actions, given the reward and punishment structure governing the entire complex of actions. Even if we observe individual sports such as tennis, badminton, or chess, the idea is to play within the rules of the game and extract the most beneficial result; gamesmanship which does not lend itself to the imposition of penalties but can tilt the result in one's favour is often resorted to.

It so happens that robots can mimic this reward-responsive behaviour of humans and actually do better than humans in this regard. Consider the boardgame *Go* in which a player is almost always faced with many alternatives in regard to possible moves: *DeepMind's AlphaGo*, a computer programme, was not only able to understand this game but defeat the world champion, Lee Sodol, in 2016. Another program, *OpenAI* was able to direct a robotic arm to solve the Rubik's cube.

In order to facilitate a better understanding of reinforcement learning and how a computer programme facilitates such learning I look at a very simple example (inspired by Rana, 2018 and similar literature). Actual applications in real life are bound to be much more complex. Consider the matrix of four squares in Figure 4.1 where the squares are labelled as 1, 2, 3, and 4. The label attached to each square is a natural number less than 5 which defines the state of the game. Given each state of the game, the AI agent (robot) can choose among four actions – "up"(U), "down"(D), "left"(L), "right"(R) – which, respectively, are associated with movements in the said direction by a single square if such movement is possible. Thus, if the robot is at 1, leftward movement is not possible and the choice of L results in it banging against the left wall of the matrix and

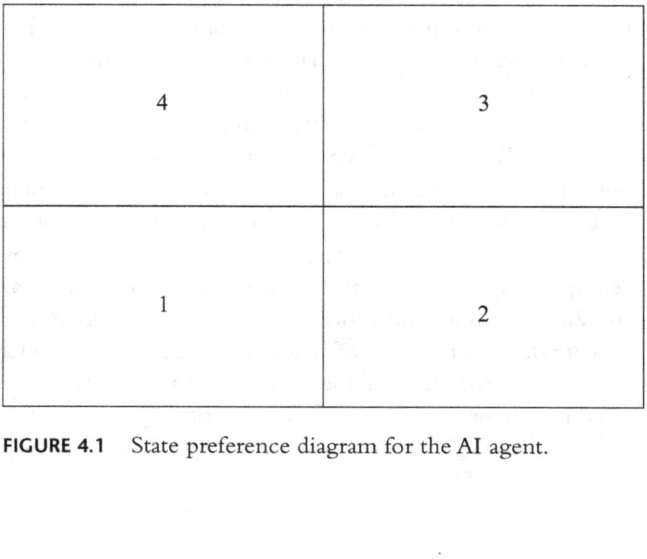

FIGURE 4.1 State preference diagram for the AI agent.

staying at 1. Similar constraints apply to movements when the robot is located at other squares.

The idea behind reinforcement learning in this scenario can be quite simple: to teach the robot to reach state 3 in the shortest number of moves (actions) from any other square. Thus, if the robot is at square 1, actions L and D are awarded the lowest reward of −1, while R and U fetch the maximum award of 1 in taking the robot to states 2 and 4, respectively. From either of these states, it just takes one correct move to reach 3, U in case the state the robot is reacting to is 2 and R in case that state is 4.

When the robot is subjected to reinforcement learning it know nothing about the correctness of moves to start with. Therefore, it starts by choosing moves at random but then tweaks its moves according to the accumulating memory of state–action–reward combinations. So, if it chooses L at state 1 (reward of −1) and then follows it up with B (−1) it stays at square 1 over an allowed episode of two moves and earns a reward of −2. In the next episode, again of two moves, it tries to do better by choosing moves other than the ones chosen in the last episode, and as the reader can see, this would result in its score improving to 2 if it chooses R followed by U or U followed by R, and 0 if R is followed by any move other than U, and U is followed by any move other than R. Thus, the reward maximizing robot learns how to enhance its scores by making mistakes and sacrificing rewards, and then using the data on mistakes and associated rewards to improve its actions. In the end it learns to move to square 3 from any other square in the swiftest possible manner, ranging from 1 to 2 steps.

The basic idea of reinforcement learning is the determination of optimal outcomes, given constraints and other rules and past evaluation of attempts to arrive at optimal outcomes. A more elaborate discussion of computer programming of reinforcement learning is conducted below but it might not surprise the reader, even at this juncture in the book, that reinforcement learning constitutes a major potential avenue for saving resources: time, material inputs, energy, and so on.

But reinforcement learning can also enable robots to perform tasks hitherto performed by skilled human labour or beyond the reach of human skills. For example, in regard to healthcare, a robot can learn to heal through reinforcement learning by looking at masses of actual records which list the following: symptoms of the patient, medications used by human doctors for treating these symptoms, and outcomes of the medication. This is achieved through reward and punishment, as embodied in a human written computer programme: a higher score as reward for the observed prescription of a more appropriate medication in response to stated symptoms. This is akin to a human sitting for successive tests with the score on each test taken to date helping her to perform better in future. Robot doctors can prove to be very useful in scenarios when there is a shortage of human doctors such as the rural areas of many developing countries.

Reinforcement learning is also ushering in a revolution in policymaking. A computer programme can use reinforcement learning to link different possible configurations of tax rates to the resulting income distributions and the associated income equality. The policymaker is thus helped in choosing the optimal tax structure for implementation.

Machine learning and economic development

A survey of the literature yields a bird's eye view of the mechanisms through which machine learning affects economic development and other parameters of economic health such as employment.

Ascertaining the suitability of tasks for machine learning

Note that, as pointed out by Autor, Levy, and Murnane (2003), every job or occupation consists of tasks, some of which correspond to higher potential for being carried out by machines or computer systems than others. With progress in machine learning (ML) some of these tasks can be taken over by computer systems. Brynjolfsson and Mitchell (2017) developed an index called "suitability for machine learning" (SML) which indicates the potential for each human task to be taken over by a machine because of ML.

In order to identify the tasks that are SML, we often need to ask ourselves the following question: what are the tasks which humans cannot directly write computer programs for but can do? (Brynjolfsson and Mitchell, 2017) Examples are face recognition or identifying different types of animals. This is because we can do these tasks but cannot delineate the rules or steps which enable us to carry these out. Thus, for example, humans can attach the same label ("name") to a face at different points of time or different labels to very similar faces at a point of time (i.e., face recognition) but cannot list out the steps which enable them to carry out such labelling. This lack of explicit knowledge of the rules/ steps implies that a computer program, where these rules are listed, cannot be written by humans. What humans can do is to identify the input and output data associated with such tasks. For example, in the case of face recognition, the input is a catalogue of labelled human faces – which computers can store in hard drives and we humans store in our brains – and an unlabelled human face. The output is the "label" we attach to the unlabelled face or the equivalent of the statement that the "unlabelled human face does not match with any faces in the catalogue". The computer can be trained on input data and output data generated through human effort. This training is used by the computer to discover, on the basis of an algorithm, the functional relationships which when applied to the input give the same output as a human.

Another example of machine learning involves getting the computer to act as a "learning apprentice" (Brynjolfsson and Mitchell, 2017): the computer acts as an assistant to a team of human consultants, a position which gives it an opportunity to witness human decision-making and tasking. Data on human decisions, under different circumstances, thus generated from observing various human consultants enables the computer to acquire a skill level which surpasses the skill level of any single consultant. It might be the case that the computer is witness to suboptimal decision-making or tasking by humans. The computer can compensate for such suboptimality by referring to a neutral mass of trustworthy information. For example, medical diagnosis based on images can be assessed for quality on the basis of subsequent medical tests and the computer can choose to learn mainly from diagnoses proven to be of "high quality" by tests.

Apart from human tasks possessing the two mentioned characteristics – humans not being able to list the sequence of steps which leads to the completion of a task and human tasks providing an opportunity to computers to learn from the humans they assist – Brynjolfsson and Mitchell (2017) list eight characteristics of tasks which lead to significant SML:

1. The task involves mapping a function from well-defined input data to well-defined output data: examples include the input and output data listed in our discussion of face recognition and the labelling of fruits, vegetables, and animals.
2. Large data sets consisting of input–output pairs for training the computer system in the performance of the task exist or can be created: there is often a high potential for observing human decision-making in circumstances when the inputs and outputs can clearly be identified – the decision in these cases, which is a product of cognition as well as judgement, is the output. Care must be taken not to imbibe the biases in human decision-making: for example, the guardians of the law (police, judges, etc.) might be biased in their dealings with individuals of different racial backgrounds and a computer system learning from data on these decisions and relevant background information might pick up these biases.
3. If the data (combination of input and output data) used for training the computer system in a task is affected by the influence of human biases in decision-making on the output data, the computer system should be able to access a gold standard for labelling observations in the data set according to the human bias displayed, and thus learn from observations where the bias is absent or minimal.
4. The task does not involve long chains of reasoning or is not based on common sense or background knowledge that is very difficult or impossible to capture in a data set.
5. If the task, when carried out by a computer system, involves a large number of functional transformations in going from the input to output, the computer system will do a poor job of explaining its decisions; the task in this case will be a candidate for SML only if such explanations are not needed.
6. There should be tolerance for some error as error free performance of ML-driven computer systems is not possible. Speech and object recognition and clinical diagnosis by a computer system learnt through ML are all characterized by the presence of some errors. As the best human experts

also make such errors, the replacement or supplementation of humans by such computer systems is quite acceptable, given that such computer systems outperform humans in regard to speed of completion of tasks and cost efficiency and the incidence of error can be made to match that exhibited by human experts through adequate training of computer systems.

7. Changes in the distribution of input data on which a task is to be performed implies that the performance of the task by the computer system on the new input data has to be preceded by retraining on data which reflects the change in distribution; tasks should be selected for performance by computer systems only if such training is possible. This possibility depends on how fast the distribution changes and how fast new training data sets can be built to mirror such changes in distribution.

8. The task should not involve too much physical dexterity and manipulability on the part of a robot as ML and available human capital for building robots has of now not been able to endow a robot with such qualities. Consider, for example, the tasks carried out by cleaning ladies in hotel rooms which involve walking, touching, bending, grabbing, turning, pouring, twisting with fingers, tearing, etc. It has not been possible to build robots which perform the entire gamut of tasks accomplished by cleaning ladies or even a majority of these tasks, though a few of the jobs such as vacuuming have been robotized.

The impact of machine learning on productivity and international trade: a discussion of channels

A study of the U.S. economy by Brynjolfsson, Mitchell, and Rock (2018) tells us that most jobs/occupations, which are basically bundles of tasks, have at least some tasks which correspond to positive and significant SML. However, not all the tasks in a job/occupation correspond to such SML. Further they conclude that unleashing ML potential for productivity enhancement will require significant redesign of the task content of jobs, with SML and non-SML tasks within occupations being unbundled and non–SML tasks across occupations/jobs in a production process being rebundled to form a new set of occupations/jobs to replace the old ones. It seems that this productivity enhancement would occur through human workers specializing in a smaller number of tasks than before due to ML, and complementarities between the increased efficiency in performance of a task taken over by machines and other linked tasks which would continue to be carried out by humans. Note that the mentioned "increased efficiency" comes about due to various factors including (i) the computer replacing the human in performance of a task, being able to perform the task more efficiently immediately after replacement; (ii) improvement in machine learning algorithms; and (iii) computer systems becoming more accurate over time as the data sets used for training become larger.

Efficiency increases in production processes, often augmented by unbundling and rebundling of tasks, are only one of the mechanisms through which machine learning can lead to economic development. One of the major transaction costs of trade between two nations that communicate in different languages is that involved in translation of material on product specifications and terms of sale. As translation becomes easier and quicker to facilitate because of the availability of machine translation facilities, the volume of trade increases. As machine translation becomes more accurate due to progress in machine learning, the hesitation of one country to buy from the other country due to the fear of not comprehending the product specifications accurately diminishes. This means that progress in machine learning should result in an increase in the volume of trade as pointed out by Brynjolfsson, Hui, and Liu (2018). The empirical exercise they carry out is that with regard to international trade resulting from purchases on an online platform called eBay. Only eBay sellers and buyers from various countries are allowed to carry out transactions. Whenever the buyer and seller are from different countries the resulting transaction adds to world exports. What Brynjolfsson, Hui, and Liu (2018) do is to consider the impact of an improvement in machine translation on international trade mediated by eBay.

In 2014, eBay introduced eMT, a machine translation system to replace an older machine translation system called the *Bing Translator* in the performance of the following task: translating queries for items made by customers in their native language into English, matching them with item listings in English on eBay's own website, and translating such item listings into the mentioned native language for perusal by customers. A group of human experts when asked to compare the quality of translation by eMT and *Bing Translator* ruled that eMT got 91.4% of its translations correct as opposed to 84.4% in the case of *Bing Translator*. The percentage of translations done by a computer system that is deemed as correct by human experts is called the human acceptance rate (HAR). Thus, HAR in the case of eMT exceeded that for *Bing Translator* by 7 percentage points or 8.3%. This is a significant enough increase, and the authors find that it increased international trade on the platform by 17.5%. Increased international trade would obviously increase the market for high quality and cost-efficient products while diminishing the market for products exhibiting low quality and low cost efficiency which would ultimately be competed out. An increase in sales of high quality and cost-efficient products would be associated with an increase in the scale of their production which is mostly associated with decline in cost and price. The beneficial impact on consumer welfare levels would be significant.

The impact of machine learning on employment

What happens to employment when machine learning results in many production tasks, hitherto being undertaken by humans, being taken over by computer systems? The answer to this question is obviously linked to the impact of ML on the scale of output, an economic indicator which is important in itself, as employment is the product of this scale and the average labour intensity of output. First consider the scenario in which the mentioned takeover by computer systems of certain tasks is not accompanied by any change in the content and number of production tasks. For a given level of output this would obviously result in a decline in aggregate time for which humans would be employed, i.e., greater unemployment. Developments in this regard might actually become very nuanced as mentioned: employers might unbundle the tasks corresponding to each job or human occupation, as previously defined, and then aggregate across such jobs to form new human jobs (Brynjolfsson, Mitchell, and Rock, 2018). This would require some training of human workers in tasks they have not done before. The firms employing them would incur some training costs which would tend to depress profits in the short run. The average number of tasks per human job and/or the number of human jobs (occupations) would shrink as machine learning takes over a significant and growing share of the number of tasks in the economy. If humans become better at the tasks allocated to them through specialization in a reduced number of tasks and/or training and so do computer systems because of progress in machine learning, the scale of output in equilibrium increases because of cost reductions. This increase in scale counteracts the effect of a diminution in the number of human tasks and reduces unemployment as we go from the short run to the long run following machines taking over certain human tasks. In the long run, the total effect of machine learning on employment might not be negative at all: a reduction in the number of human tasks might be more than compensated by an increase in the scale of output of goods and services and an increase in the man days for which a typical human task is performed.

However, this is not all. Machine learning, as mentioned, might enable computer systems to perform tasks never ever performed before by humans or computers on an economic basis. When these are complemented by human tasks, new or performed in the pre-ML era, it might become possible to profitably produce new products, processes, or services (Brynjolfsson and Mitchell, 2017). While some of these might be produced within the ambit of existing industries it is highly probable that new industries might sprout up. This would again have a positive impact on employment in man days. Examples consist of subscriptions to movies and music videos sold on the internet: these subscriptions not only provide the consumers with a wide choice but also are able to customize offerings to millions of consumers on the basis of revealed choices. Note that such customization through human labour is both slow and not possible for more than few consumers at a time.

Note that ML-generated cost reductions in regard to existing products would lower prices of these products and free up purchasing power. The liberated purchasing power can then be used to augment the scale of purchases of existing products and consume new products. As demand (purchases) goes up there is an increase in quantity supplied to meet this increase and thus an increase in the level of employment. To summarize, use of applications of ML in production processes all over the economy constitutes an innovation on the supply side which results in cost and price decreases, thereby increasing purchasing power and stimulating supply side responses in the form of an increase in output/employment.

Brynjolfsson and Mitchell (2017) explain the impact of ML on employment on the basis of computer systems created through ML substituting for humans in the performance of certain tasks, and the resulting changes in the scale of employment of humans in tasks that remain unautomated. First, automation of a task due to ML inevitably results in a decline in the price that the entrepreneur has to pay for the task and consequently an increase in the quantity of that task demanded by the entrepreneur as well as quantities of other tasks that remain unautomated and carried out by humans but are complementary to the automated task. The increase in quantity of the automated task as a result of price decline obviously depends on the price elasticity of demand of the task (a measure of the responsiveness of quantity demanded by the entrepreneur to price) while an increase in the demand for the mentioned complementary tasks that follow such automation will inevitably result in an increase in human employment in these complementary tasks as well. Lower the price elasticity of supply (a measure of responsiveness of quantity supplied to price) of human labour for the unautomated task lower is the increase in employment of humans in complementary tasks: additional labour that can be hired by a typical entrepreneur to perform the unautomated tasks will be difficult to come by even if a higher wage is paid. To summarize, both the increase in quantity of the automated task and that of each human task that is complementary to it will depend on the price elasticity of demand of the automated task (derived from price elasticity of demand of the product that the automated task helps to produce) and the price elasticity of supply in regard to the unautomated (human performed) tasks.

To comprehend this better, we may consider an entrepreneur who produces a unique product and thus caters to the entire market for it. Now assume that some of the tasks in the production process become automated and cheaper due to ML. As production becomes cheaper, it gives room for the entrepreneur to lower price, increase sales by attracting more customers, and enhance profits. Thus, if price elasticity of demand prior to the mentioned automation is small, it implies that sales enhancement in response to automation-enabled price decline, which exhausts the room created by reduced cost, will be small. This in turn will imply that the quantity undertaken of the task subjected to ML, and by implication the unautomated tasks complementary to it, will not go up much after automation. As mentioned, the price elasticity of supply facing the entrepreneur in regard to human capital is also important: the entrepreneur cannot cash in on ML gainfully by producing substantially more of the product, if this would involve hiring human capital at a considerably elevated price to do tasks that are complementary to the task automated by ML.

Brynjolfsson and Mitchell (2017) have also commented on how substitution of humans by computer systems and robots has changed over time in the AI era.

This can obviously be split up into two eras – the pre-ML IT era and the post-ML IT era. In the pre-ML IT era, computer systems took over routine, repetitive, and structured tasks hitherto performed by human workers who belonged to the middle of the wage and skill spectrum. These were tasks which involved explicit steps by humans that the humans were aware of and human programmers could build into a software. Examples of jobs impacted included clerks and factory workers. Because many of the tasks performed by such employees were taken over by computers, the employment provided by such jobs tended to decrease. Jobs which were not impacted included (a) those which were associated with physical labour and required dexterity and manipulability and (b) others which required a high level of skill and fetched a high wage because of the required high levels of human capital and judgement or emotional intelligence. The former type of jobs belonged to the bottom of the wage spectrum and the latter mostly belonged to the top.

In the post-ML IT era, many more jobs have been impacted by computerization. These include jobs consisting of some tasks for which humans have not been able to articulate stepwise procedures (in other words, explicit strategies) but associated with input and output data lending itself to specification of an explicit functional relationship between input and output that mimics output generation by the human employee. For example, a computer system can take over the task of compilation and classification of documents for a legal case, a phenomenon which would have a direct depressing effect on the size of legal teams. However, it cannot perform the task of interviewing human witnesses well as this task requires emotional intelligence (i.e., identifying human emotions and relating these to proceedings to date), common sense, and extensive background knowledge. For the same reasons, it cannot come up with a winning legal strategy. Similarly, computer systems can take over the task of reading medical images and using these to come up with medical diagnoses; however, these would be far worse than a human doctor in interacting with other human doctors or communicating with and comforting human patients. Some trimming of medical teams might therefore be possible through ML-led computerization.

How reinforcement learning facilitates optimization in production and consumption

The consumption problem: helping humans to move from satisficing to optimizing

Human beings are often said to have a tendency to optimize, given that they are rational organisms. Consider a household which wants to consume a bundle of commodities after purchasing the various constituent items. The items can be purchased in various ways given that there are various outlets selling items: virtual outlets as well as physical stores. Each physical store is characterized by availability of some of the items at stated prices, the possibility of non-availability of others, and the need for travel by the household to the store; each online store is again characterized by prices for the items which are available and possible non-availability of others, with the household being able to access these stores at zero travel cost. Thus, the household has data regarding prices, availability/ non-availability of individual items, the distance that has to be travelled to each store, and the time needed for travel as well as monetary cost of unit distance travelled, given the price of fuel and the average speed of travel. If human beings actually do optimize as is often claimed, each household will choose to distribute its purchases of the commodities that make up the entire targeted bundle over various stores so as to minimize costs.

This tendency to optimize in the face of incentives and disincentives, as well as constraints is supposedly the hallmark of human behaviour. Textbook treatments of human choice, for example, in economics using mathematical tools, devote a lot of space to the process as well as the implications of optimization. However, humans are faced by constraints in regard to time, given that acquisition of information and its processing or computation, all needed to generate an optimal plan, are time-intensive. Moreover, the human brain is simply unable to undertake computational exercises involving data inputs exceeding a certain critical volume and constraints in excess of a critical number. For example, a human being trying to compute the value of $2^{10} + 3^{10}$ with a pen and paper displays an efficiency, in terms of proportion of the task completed per minute, which is probably less than 2% of a pocket scientific calculator. If the same human wanted to find out the sum of the tenth powers of the first 100 positive integers even "punching on the calculator" would be a bad idea and a software programme would be the best remedy. The basic idea being conveyed here is that though we are an intelligent species, the ability of our brains to perform computations is extremely limited. Thus, such computation is time-intensive and complex computation is not possible.

But how do we then solve the problem of "optimal choice" if as a species our ability to compute is limited? Traditionally we have not been equipped with computers and calculators; even pocket calculators came into use in households in the middle of the 20th century (Akanegbu and Ribeiro, 2012). The problem of "optimal purchase" can be extremely complicated if our reach includes hundreds of stores, each characterized by different availabilities of items and prices and located at various distances from each other and the concerned household. Given that humans are both computationally challenged and have to battle constraints in regard to time, Nobel laureate Herbert Simon suggested that human beings try to obtain satisfactory solutions to complicated problems of choice rather than the optimal solutions. This obviously includes purchase of bundles of consumer goods by households.

Thus, for a household, there might be hundreds of ways to purchase a given bundle of commodities – containing positive amounts of a finite number of items – by buying from various stores. And there might be millions of possible bundles which can potentially exhaust the budgeted consumption expenditure, with satisfaction of the household after consumption varying across these bundles.

For example, given the obvious challenges posed by human limitations in regard to computational capacity and the paucity of time – consider the act of grocery shopping which is allocated an hour or so at the end of a work-laden week – most humans in an American city look for a commodity bundle purchased from one specific supermarket or an online retailer, with a reputation for reasonable prices, which meets all their needs to satisfactory extents. Thus, what humans do in real life is to try and get a satisfactory result, i.e., purchase a good enough commodity bundle at a reasonable cost which is less than the budgeted amount. In doing so the household takes on an "information burden" which it can handle. Obtaining a satisfactory solution, which is not necessarily the optimal solution is known as "satisficing", a term coined by Nobel laureate Herbert Simon (1956) and based on his work in the 1950s and 1960s (Simon, 1955 for example). Quoting Simon (1956), "evidently organisms adapt well enough to 'satisfice'; they do not in general 'optimize'". Prior to the emergence of AI, even giant companies obviously looked to base their purchases using "satisficing behaviour".

As humans, we do appreciate the "optimum" as a concept and as something worth striving for. This appreciation is reflected by the considerable space devoted to optimization in textbooks on mathematics and economics. However, given our own limitations, as discussed in the last paragraph, in real life we often compromise to "satisfice". Simon's conclusions can be qualified to say that if humans can optimize quickly, they do. If I am faced with two alternatives to move from location A to B, other things remaining the same, I will choose the alternative that results in travel along a 40 km stretch of road at 40 km/hour (thus lasting 60 minutes) over that associated with travel along another stretch of 60 km at 48 km/hour (75 minutes). Any person with a school education can solve this problem in a couple of minutes and proceed to make the optimal choice. All of us solve – and have historically solved – hundreds of problems of optimal choice such as this as we journey through life.

However, if optimization is time-consuming for the human brain, humans have traditionally satisficed because time itself is scarce and therefore valued by humans. In such cases, it is "optimal" to not "optimize" but "satisfice". In other cases, optimization might not have been traditionally possible before the advent of AI in our lives because the human brain can only process a limited amount of information. The problem of the consumer choosing among the bundles of commodities that are supported by budgeted expenditure is one of them: in a world in which there are hundreds of consumer goods on sale the number of such bundles can easily run into millions. No one has the time or perhaps even the capacity to rank these bundles and then choose the optimal one; therefore, a typical human even in this AI age picks a bundle which is satisfactory for purchase instead of trying to reach the pinnacle of satisfaction.

The robot, a recent phenomenon with computational speeds and capacities millions of times that of a human, can easily be trained to be precise and therefore to "optimize" instead of "satisfice". In contemporary times, the act of "optimizing" can be outsourced to the robot at very little cost through a reinforcement learning programme written by humans. AI is now being used to great advantage by firms in procurement (Sievo website and Planergy website), including sourcing of outputs from potential suppliers. There is no reason why such technology cannot be used by households to optimize their purchases. Discussed below is an example of how a robot can be taught through a reinforcement learning programme to make purchases for the household.

Consider, the task of buying a single bundle consisting of 12 oranges and 4 bananas from one or more out of three alternative stores, A, B, and C. Store A sells only bananas at Rs 4 per piece; Store C sells oranges at Rs 4 a piece; and Store B sells oranges as well as bananas at Rs 3 per piece. However, Store A and Store C are online stores and shopping at these stores requires no travel costs. On the other hand, travel to and from Store B, a brick-and-mortar store, requires Rs 20.

The robot being a precise instrument of choice can consider all possible ways of buying the targeted bundle. It is easy to see that satisfying the entire demand for bananas and oranges from the two online stores involves an expenditure of Rs 64. Buying the same entirely from the Store B involves spending Rs. 68. Apart from these, there are ways of satisfying the stated demand by buying from Store B and one or more of the other stores, but this is clearly an even worse solution as once a person has decided to go to Store B she should buy everything from there, given the lower price of oranges as well as bananas in that store compared to the online stores.

The way the robot (AI agent) can be taught to minimize costs given the various constraints is to just ask it to choose among the various ways of buying the mentioned consumption bundle of 12 oranges and 4 bananas. Each choice is then rewarded through a "reward function" which awards a score to the choice ranging from, say, −100 to 100. Any score short of 100 for a choice implies scope for improvement; the further the reward is from 100, greater is the scope for improvement, with a score of −100 implying highest possible scope for improvement. When the robot arrives at a choice, it is able to determine from the "reward score" how it has performed in regard to optimization. Thus, the "reward score" calibrates the performance of the robot, and the information on past performances helps the robot to fine-tune its future performance in regard to choice; in this specific case, one can think of various rounds of choices with each round corresponding to new numbers for prices and travel cost. As the rounds goes by, the robot learns from the feedback to its performance through the reward scores and adjusts its choice in the direction of cost minimization.

Rewarding performances with appropriate scores provides enough data for improvement of performance. The process can go on until perfection or near perfection in performance is attained consistently. This is quite akin to a child learning to walk on its own − each stumble or fall provides information to the child in its attempt to walk better. For example, if she takes a rather large step and falls, the negative reward in the "guise" of a fall teaches her to take smaller steps. This feedback loop of behaviour-reward-behaviour, where behaviour earns an appropriate reward and the latter in turn is a determinant of behaviour, constitutes the kernel of reinforcement learning in humans. Moreover, as we have seen, this kernel is retained in human developing reinforcement learning programmes for robots/computers.

Once the robot grabs the principle behind cost minimization of purchases it can apply it to any kind of "purchase problem". It is quite possible that optimization made possible by employing a robot yields gains of 10%–20% over the manual use of satisficing; after all, purchases obtained through "satisficing" often involves examining only a fraction of the total number of possible choices to arrive at a satisfactory choice.

The economic implications of outsourcing optimization in purchase to robots are significant. Given that robot butlers such as Siri, Alexa, Cortana, and Watson are available almost free of cost, one would expect the mentioned "optimization devices" to cost very little, but as mentioned, these can help to slash household expenditures. Giant stores and supermarkets using large budgets for advertisement and other kinds of sales promotion to lure consumers might have to look more carefully at the prices they charge. It is also possible, in fact highly probable, that such AI devices can combine optimization with the procurement of the optimal choice.

Think of the following situation not very far into the future: you give your robot butler a budget and a shopping list at the beginning of every month. From thereon she can take over and purchase all the items in the list at the lowest cost. This is a big jump over where most of us are now. But further advances are possible if the robot also knows your preferences and has an exact picture of the inventory of items at home: it can actually come up with the shopping list. Below I discuss reinforcement learning that helps platforms such as YouTube and Netflix continuously enhance the knowledge about our preferences; in fact, they might know more than what we do about our own preferences.

Tailoring the product to the consumer's preferences: the YouTube model

In economics, the theory of revealed preference (Roper, Britannica Website) is a formal but useful statement of the obvious: what a person chooses for given income and prices informs us about his preferences. But even if we take a single product – say movies, which are characterized by difference in genre (action, romantic comedies, period dramas, horror, etc), linguistic medium used, lead actors, style resulting from the tastes of the director/producer – an immense amount of information would be needed to determine a consumer's preferences: a ranking of movies of various types. Given that each viewer has different preferences, a lot of information on actual choice would have to be collected at the level of each viewer and then processed. Also note that our preferences are time- and context-dependent: a couple might like watching romantic comedies on Friday night to celebrate their togetherness after a hectic work at the office, run all the required chores on a Saturday, and then prefer to watch a horror movie to break the monotony of a Sunday afternoon. In other words, a deep knowledge of a person's preferences has to be built on a mountain of information and facilitated through accurate processing.

It would be impossible for a team of humans, let alone a single person, to generate this knowledge for millions of viewers but companies such as YouTube with a viewership in excess of 2.5 billion (Newberry, 2021) and Netflix, with more than 200 million subscribers (Statista website, 2021) are doing precisely this successfully for highly differentiated products such as movies/videos by using reinforcement learning to train its AI agents. The YouTube AI agent does two things: when a person visits the website it generates a list of suggestions, the most recommended videos on top followed by those which are lower in the pecking order in regard to being recommended; it also makes suggestions in response to a search by the customer encapsulated in a few words. For a first-time viewer, the AI agent has no knowledge regarding her preferences on which it can base its recommendations but as the customer chooses from among the recommended videos, watching some of these fully while aborting its viewing of others, she offers knowledge regarding her preferences to YouTube. The more she chooses and views, deeper is this knowledge base. YouTube in recent times has been quite transparent about the AI techniques it uses to generate this knowledge (white paper written by Covington et al., 2016 and YouTube Creator Academy Website) as it inculcates a belief among customers that the company is responsive to its preferences and needs This enables YouTube to maximize the viewing of videos by each of its customers which in turn has beneficial implications for its revenues from advertisements, acknowledged by YouTube itself to be its main source of income (YouTube website), given that such revenue is positively impacted by the total number of person-hours of viewing on YouTube. To clarify, 1 person hour of viewing results from a single person watching for 1 hour and when we add the number of hours viewed across all viewers in a year we arrive at the total number of person-hours viewed, which as of now is more than 365 billion (Newberry, 2021).

If companies such as YouTube can use their deep knowledge of our preferences to recommend movies and videos that suit our tastes, many of which we are not aware off, the day may not be far off when robot butlers would be able to draw up a list of household purchases, combining their intimate knowledge of household inventories and the preferences of household members. They would then be able to use their powers of optimization to minimize the cost of purchase of the list drawn up, as discussed.

Optimizing energy consumption: Google

Google's services, such as Google Search, Gmail, and YouTube, are facilitated by thousands of servers housed in its data centres. The infrastructure governing these data centres contain built-in checks for reliability, efficiency, and safety, many of which are driven by AI. The operation of the information technology equipment inside the data centre produces heat which has to be reduced through cooling (Heslin, 2015); however, various actions such as the use of cooling material or getting nature to manufacture free cooling agents can be taken to bring about the required cooling (for details see Gamble and Gao, 2018). Cloud based AI is used to generate snapshots of the data centre cooling system every five minutes. These are then fed into deep neural networks, which consider various possible actions to bring about required cooling to choose, to the best of their ability, the action which minimizes energy expenditure subject to safety constraints. A round of checks follows before implementation of the chosen action. The AI system has been able to achieve 30% energy savings to date; given that reinforcement learning is all about learning from performance, a higher energy saving is predicted in the future. Such use of AI can not only result in energy savings but become a powerful tool for checking local pollution as well as climate change.

Our exploration of reinforcement learning tells us that it can help consumers minimize expenditure for planned purchases or equivalently consume more for a given budgeted expenditure. Such minimization will make the relative magnitudes of purchase from different outlets much more sensitive to the prices they charge than that in the era preceding the use of reinforcement learning. This in turn implies that price competition will increase and consequently the importance of commercial heft exercised through large advertisement budgets might recede. Procurement by firms using reinforcement learning can help them to economize on inputs such as energy and materials as well as spend less to purchase the input bundle required for production. This would help to reduce the cost of production. To summarize, the impact of reinforcement learning on consumption practices would force suppliers to compete more in terms of prices while the impact on production practices will allow them to actually lower prices. In short, we are looking at more "price aware" consumers, with enhanced price awareness facilitated by AI, forcing more cost reduction by firms and eventually greater price reduction.

Greater price awareness on the part of consumers and more price reduction, products of the use of reinforcement learning in regard to consumption and production decisions, both would contribute to lower consumption expenditure for the same consumption bundle or greater consumption for given income. As mentioned, there would be salutary impacts for local pollution, global warming, and the use of scarce material inputs. Consumer welfare would be clearly benefitted. Note that use of reinforcement learning applications by consumers to optimize purchases will not displace people from jobs as satisficing is presently performed by consumers on their own. In regard to firms, optimization of operations, a result of optimization in procurement as well as use of inputs, might actually involve displacement of employees previously involved in "satisficing" in regard to these operations. But at the same time, the mentioned optimization for both consumers and firms will be associated with generation of jobs for writing and running programmes and monitoring their performance.

But reinforcement learning programmes might display human employees from other positions. Below I look at how these programmes are being used to pilot driver-less cars, thus threatening to make human drivers redundant and to manoeuvre vehicles such as rockets in space, probably without the same adverse effect on human employment.

Use of reinforcement learning to develop self-driving cars

Through reinforcement learning, the robot (AI agent) learns how to pilot a self-driving car. The robot is just made to manoeuvre the car inside a simulator which, just like a video game, mimics the real world the driver has to face: red lights, "go slow" signs, other cars whizzing by or past, and so on. There are negative rewards for wrong actions (Cohen, 2019): driving past a red light, ignoring a "go slow" sign, or driving the car onto the simulated pavement or into another car. The reward maximizing pursuit adopted by the robot implies that the actual experience of getting negative rewards for wrong (inappropriate) actions will lead to its eschewing these actions in the future.

Once the robot has learned how to drive inside a simulator it can be put to test in actual traffic, with a human driver always ready to take control of the wheel in an emergency (see Mindy Support website). The robot is assisted by cameras which take pictures of the surroundings characterized by objects such as cars and pedestrians; radar detectors send pulses of waves which rebound from nearby objects and help calculate their distance from the car as well as the speed with which they are approaching the car; lidar sensors supplement the work of the cameras in generating a three-dimensional view of the surroundings. The effectiveness of self-driving cars in finding their way through a sea of objects without colliding or violating traffic rules therefore depends on the quality of the reinforcement learning programme to which the AI agent is subjected as well as the assistance it gets from objects such as cameras, radar detectors, and lidar sensors.

It must be remembered though that driver-less vehicles are at the extreme end of the spectrum of autonomous vehicles. Usually, five levels of such vehicles are considered with AI able to exercise increasing amounts of autonomy for a higher level of the autonomous vehicle, i.e., perform a greater proportion of the actions used to control the car – for example, steering the vehicle, braking, and accelerating. In all of levels 1–4, drivers are required to intervene under conditions of an emergency, though for a level of 4, under certain ideal conditions the human driver can take a clean break from the task of managing the car and even take a nap. In levels 2–3, the human driver is also assisted by AI through facilities such as collision detection, lane departure warning, and adaptive cruise control, which adjusts the speed of the car to traffic conditions and the speed limit. It is the level 5 vehicles which are fully autonomous – these can be used as fully robotised taxis to deliver food, etc. (Kapoor, 2020).

Level 5 and level 4 technologies are not fully prepared for commercial exploitation as these are still susceptible to cyber-attacks (Research and Markets Report). Moreover, there are some legal issues in regard to all types of autonomous vehicles: for example, if there is an accident there might be problems apportioning the fault between the AI system and the human driver which it assists; it can become very difficult to distinguish between safe and risky driving, given standard responses of the AI to different situations as opposed to the use of human reflexes in driving and averting accidents, with this distinction key to handling issues related to insurance; finally, in case of level 4, it is very difficult to determine the time at which it is legally permissible to hand over the control of the car totally to AI.

These are grave issues. Adoption of level 4 and 5 technologies through legislative consent thus becomes very risky for legislators: critics and the public might judge them very harshly if there is a massive accident, involving one or more autonomous vehicles, resulting in significant mortality.

But market surveys show that the lower levels of technologies for autonomous vehicles are becoming extremely popular in spite of possible legal problems. As a result, the autonomous car market, which was worth around $21 billion in 2020 is poised to reach around $62 billion in 2026, expanding in 2021–2026 at an annual growth exceeding 22% (Research and Markets Report). Over the past 5 years, $50 billion have been invested into the autonomous vehicle industry with 70% coming from outside it, with the government a key player because of anticipated social and economic benefits.

The hub of the autonomous vehicle industry lies in North America as technology giants such as Google and Tesla are headquartered here. But the Asia-Pacific region – China, Japan, and South Korea and even emerging economies such as India – is showing major activity in regard to autonomous vehicles. For example, in India, automakers and research institutes have been entering into collaborative agreements for exploration in technologies for autonomous vehicles (Research and Markets Report): for example, that between MG Motor India and Indian Institute of Technology, Delhi for research in electric and autonomous vehicles in March, 2021.

It is interesting to speculate about what would happen in countries such as India and Bangladesh where driving is often a very strenuous exercise, given the high volume of traffic and its unruly and heterogenous nature, and therefore there is a huge demand for chauffeurs not only from companies but from well-heeled individuals in their private capacity (Awal, 2011). With the middle and upper classes booming on account of rapid economic growth, one would expect the demand for chauffeurs to increase over time resulting in employment opportunities that spread wealth. The spread of autonomous vehicles can be a major dampener in this regard if it decreases the tedium and stress associated with driving, especially as people switch to higher and higher levels of autonomous vehicles. In short, an increasing proportion of vehicle owners might prefer to drive their own cars, assisted by AI. There obviously is a possibility that the employment opportunities emanating from this source will start contracting though that is not an imminent danger.

Use of reinforcement learning in rocket science

Consider enabling an AI agent to land a rocket through reinforcement learning. The state of the environment is captured by data in regard to, say, eight variables and the AI agent responds to the state of the environment, thus captured, by choosing any one from a limited number of possible actions. The programme for reinforcement learning assigns a "reward score" which conveys the appropriateness of the action taken to the AI agent (see Gupta, 2019). Each episode comes to an end when the rocket crashes, fetching a reward score of $-A$ points, or comes to a rest after landing, fetching a reward score of A points. As the reader can gather, the idea is to teach the AI agent to land the rocket, without it crashing, in the shortest possible time, getting a score of A for such landing and minimizing the sum of the negative scores earned for inappropriate actions during the period of operation.

Conclusion

Thus, we see that the use of machine learning marks a quantum leap in robotization as it greatly enhances the range of tasks which a computer can perform in a cost-efficient manner. Therefore, the proportion of existing tasks related to production that can be automated in a cost-efficient manner increases and new tasks leading to new products are also undertaken by machines. There are beneficial implications for human welfare in the form of enhancement of product variety and price decrease induced by cost decline. The implications for employment are mixed: increase in output induced by cost reduction and an increase in product variety would tend to boost human employment but a narrowing of the range of existing tasks performed by humans can have exactly the opposite impact.

5

THE IMPACT OF ROBOTIZATION ON THE ECONOMY AND THE LABOUR MARKET

A look at theoretical and empirical studies

In this chapter I look at the impact of robotization on (i) gross domestic product per capita and human well-being and (ii) wage and employment outcomes in labour markets both through substitution of human labour and enhancement of productivity of human employees. I interpret "robotization" very broadly here as the varied use of artificial intelligence in production and consumption: the use of software and hardware to consolidate and collect information and process it to render services of various types which generate value for humans; the use of the internet to search for information, form networks, and to undertake transactions which had previously often been associated with prohibitive transaction costs; and the use of industrial robots to substitute profitably for manual work in the production process. Note that (i) is not independent of (ii), though such linkages have not been given due importance: wage and employment outcomes have implications for the income distribution and, because of the influence of household income on the propensity to consume, for aggregate consumption and savings and therefore economic growth.

How the two prophets, Schumpeter and Keynes, viewed the combined future impact of capitalism and technological progress

Schumpeter on how capitalism evolves in a direction which leads to betterment of human well-being

Before I start this sub-section, we have to realize that our analysis of the impact of robotization has to take into account the fact that "robotization" for the most part is taking place in capitalist economic systems. Capitalism, as Schumpeter (1943) pointed out, is an evolving system not just because of changes in the

DOI: 10.4324/9780429340611-5

social and political environment, by way of revolutions, wars, etc., but primarily because of its internal dynamics: as the capitalist system grinds on it generates new technologies, new markets, new products, and new sources of supply of factor inputs. There is a tendency to judge capitalism at a point of time on the basis of how monopolistic tendencies are seen to be stifling price competition among firms, but such static analysis totally ignores that it is capitalism's evolution over time and how it makes the necessities of life available to a greater proportion of the population that should be the focus of any barometer of the desirability of the system. Large firms dominating an industry might not indulge in significant and explicit price competition, but they indulge in quality competition through upgradation of production processes, development of new products, and discovery of new markets. Moreover, these avenues are not available to small firms. As production costs tumble in absolute terms or relative to per capita income, these result in irreversible price decrease, with labour hours needed to produce vital consumer durables (televisions, laptops, etc.) and other consumer goods decreasing over time and ushering in higher standards of living for the masses. It is important to recognize that in this age robots would indeed be employed mostly by large firms to bring about secular cost and price decreases as well as a major revision of the scale of output, with small firms losing out, but as Schumpeter indicated, this might indeed be a means for the "creative destruction" of capitalism to evolve and cater to a larger proportion of the population. What Schumpeter did not however point out is that the evolution of capitalism is often co-mingled with the efforts of those who shape economic and industrial policy to curb its excesses; consider, for example, the various legislations passed to curb the competitive instincts of entrepreneurs during the early Industrial Revolution in Britain or the Anti-Trust Laws in various capitalist economies to discipline the use of entrepreneurial muscle to eliminate smaller but efficient firms. However, to a large extent the Schumpeterian critique of the myopic appraisal of large firms is valid; further this needs to be kept in mind when I conduct our analysis of the impact of robotization on the labour markets and the economy.

Keynes on the impact of technological progress: temporary negative effects and permanent positive impacts

From Schumpeter I move to another economist and thinker whose views are as relevant today as these were a century back, given that he devoted a lot of time to thinking about the impact of technological dynamism, a major characteristic of the current artificial intelligence (AI) age. It was John Maynard Keynes who wrote, around 90 years ago, about the threat of technological employment (Keynes, 1930). But Keynes did not think about this problem as a permanent one: seen through his lens the phenomenon of robotization can actually be expected to have a positive impact on the average well-being of the human race. If robotization is considered a landmark in the evolution of capitalism, then Schumpeter and Keynes were talking about nearly the same thing. Keynes referred to the time at which he was writing and pointed out the following problem: technology was evolving at so rapid a pace, in the direction of lower labour intensity of output, that it was proving to be impossible to absorb the labour liberated from employment through labour-saving technological progress of the dynamic sectors. Job creation induced by the lower cost of production that also resulted from technological progress was simply proving to be not enough as a sponge for carrying out such absorption.

Keynes points to the era between 1700 and 1930 as a period of remarkable growth in which technological progress led to a remarkable and several fold increase in the standard of living. He recognizes the problem of technological unemployment: enhanced sophistication of the mechanized means of production resulting in economy in the use of labour that more than overwhelms the allocation of labour to new uses. But he sees it as a temporary problem: he expects the surge in the standard of living to continue in the long run and technological progress to unravel a path which would take human civilization to a stage in which the compulsion to seek work in order to eke out means of subsistence would not exist; he sees a typical member of the generation living in the 21st century working around 15 hours a week, and that too because of human beings being bound by old habits of working. Keynes's predictions do not however seem to stand the test of data, at least universally: if we look at the U.S. economy in 1979–2011, the real entry-level hourly wage for men recently graduated from high school fell by as much as 25% according to data from the Economic Policy Institute, while the number of jobs that offered health insurance as a percentage of the number of jobs fell 32.5 percentage points from an initial level of 63.3% (Cooper, 2012). Growth in median wages in the same economy was very slow in the last three decades of the last century and negative in the first decade of this century (Lowrey, 2011). At the same time, unemployment or lack of suitable employment for those in the labour force continued to be a significant problem. Given the tendency for recent economic development to lead to greater concentration of wealth and income the world over, we should expect to find that major economies other than the United States have also been characterized by similar problems.

It is however possible that when Keynes was referring to the "long run", he had in mind a period longer than 1979–2011 and he expected the path charted out by technological progress under the influence of prudent policymaking to solve the problems mentioned in the last paragraph. It is possible that some of Keynes's predictions could come true: the shortening of the work week resulting from machines taking over from men as primary movers in the production process;[1] the ability of technologies in manufacturing and agriculture to generate so much output that everybody could be potentially lifted above subsistence, etc.[2] But at the same time, one fails to understand from his essay how exactly the phenomenon of significant unemployment resulting from the emergence of labour-saving technologies would be tackled so as to guarantee for every person his/her subsistence. There is surely an underlying assumption of governments evolving in their role as welfare states and inducing combined use of greatly improved technological machinery and redistributive mechanisms to bring about the predicted outcomes. It would therefore be wise, as has been done in the course of this and the subsequent chapter, to look at not only what technological progress in the form of robotization and a deepening of the AI revolution has to offer by way of positives in regard to levels of representative income and well-being but also to speculate about how the potential adverse effects of these developments, capable of significantly neutralizing some of the mentioned positives, can be contained through policy.

Spence's clear articulation of how artificial intelligence contributes to economic growth and enhancement of welfare

Unlike Keynes and Schumpeter who banked on their powers of foresight to predict the impact of technological progress and automation on human standards of living, Spence (2011) could use the fairly well documented history of the first 70 years of computerization to feed into his predictions regarding human well-being and income in the AI age. Starting from the era of World War II which saw the launch of the computer – initially only for administration, defence, and research – the accent was on making computers smaller, faster, and portable. This finally resulted in computers being used on a much larger scale, i.e., by households and businesses and not just by governments and scientists.

The effects of computerization took about 40 years, according to Spence, to show up in the empirical estimates of productivity. Much of the impact of computers on productivity materialized when they were used in a network: the greatly superior ability of computers, as compared to humans, to process information had a significant positive impact on the global economy when the electronic databases scattered all over the globe were almost converted into one seamless database through a network. The internet, the search engines that were used to identify relevant information from the internet and extract it, and the computers that logged onto the internet and operated its search engines corresponded to such a network: a global pool of knowledge characterized by exponential growth through interminable cycles of extraction-processing-extraction was born. Initially the internet was used only by scientists to quickly extract information whose processing formed the basis of results. These results were then published online in the form of research papers and reports and in turn became the basis for more research, more results, and yet more research. But soon it was realized that the same networks could be used to provide services which did not require the provider and the user to be in close proximity to each other. E-billing, e-government, e-learning, e-medicine, and numerous other e-services were born. Services could not only be provided but relevant documents exchanged and payments made; this in turn widened the range of services that could be profitably and conveniently provided both from the point of view of the service provider and the consumer. Retailers started operating through the internet, displaying catalogues of products which listed their characteristics and prices, and finalizing sales and shipping purchased products. Information that was relevant for making consumer purchases of products and services and for accessing government benefits/facilities were available at the click of a button and hugely reduced transaction costs; many transactions, which were earlier deemed as not generating value for the consumer because of associated costs incurred in travel and in searching for information, now became worthwhile to undertake. The reduction in transaction costs led to a massive growth in transactions and made a huge positive contribution to economic growth and welfare.

The AI revolution has generated other major benefits: it has enabled jobs to cross international boundaries and move to persons given that movement of persons to jobs through international immigration is possible only on a very limited scale. The AI revolution has thus enabled formation of a seamless labour market and knowledge economy: human capital formation is possible without the bridging of physical distances between the providers and recipients of human capital, which can then form the basis for employment during the course of which the employer and employee can remain separated by distances of thousands of miles. For example, call centres in developing countries such as India have employed citizens to provide customer services on behalf of Western companies to their consumers in Western countries on the basis of customer and product information transferred through cyber networks; students located anywhere in the world can be tutored by a qualified person without her changing location, with the internet providing information about/to potential tutors and students so that tutor–student pairs thus formed can serve as the basis for paid knowledge transfers.

Note that the internet provides a storehouse of information to the employer about potential employees, with search engines allowing them to sift through this global information pile and hire those who represent the best combination of knowledge and talent, with location of the potential employee not mattering in many cases. Entrepreneurs can also use the internet to search for the cheapest or best supplies of raw materials; this very capability enhances the profitability of production and induces the mentioned suppliers to become more competitive in regard to both quality and price. In general, competition at different layers of the supply chain is enhanced with consumers benefitting from both better quality and lower price. This serves as a basis for enhancing consumption, both in regard to its scale and variety, and thus provides an impetus to consumer welfare and well-being.

But quite apart from what has been mentioned, it is important to realize that contemplation of any human action or transaction always has to be complemented by the search for relevant information and its processing so that costs/benefits of such actions can be anticipated. The use of computerized networks has helped to immensely reduce the time for both search and processing; the salutary implications of this reduction for leisure time are significant. Travel to perform chores for registering cars or applying for passports is now not needed in many countries; students can now attend classes at home, a convenience that has kept the process of education alive during the dreaded Covid-19 pandemic. The result has not only been the mentioned saving of time but also the generation of tendencies to reduce traffic gridlocks and vehicular pollution. While it is easy to see that these outcomes have significantly elevated human welfare, it is often difficult to measure such elevation in monetary terms. Of course, it must be realized that a whole new world of cybercrime has emerged; terms such as *hacking* (unauthorized access to data in a computer or system), *phishing* (people sending emails purporting to be from reputable companies so that they can acquire personal information such as passwords and credit card numbers), and *snooping* (witnessing online activities of others without their permission and knowledge) have become common in the vocabulary. Even major crimes against humanity such as terrorism have been given a boost by recently emerged cyber technologies. In summary, computerization and consequent formation of networks has resulted in a reduction in the transaction costs for many crimes; this has obviously given a boost to criminal activity which has detracted from the mentioned important positive contributions to welfare.

The discussion in the last couple of pages is based on the contributions of Spence. As mentioned, Spence's projection of the future is more concrete than those of Keynes and Schumpeter made in the distant past, based as it was on a very clear understanding of the impetus provided by the development of computers for the formation of networks and informed decision-making, the volume of transactions, and thus economic growth. Yet even Spence did not ponder much over the fact that the computers and their networks enabled through the internet resulted in the saving of massive amounts of labour: in many aspects of production, labour used for information searches and processing gave way totally to computers; robots replaced manual labour in agriculture, manufacturing, and retail; and finally, new products such as online subscriptions to movies and music replaced rentals involving face-to-face interactions on the shop floor. Clearly, there is the possibility of a huge depressing effect on employment and through it on demand faced by entrepreneurs. As discussed earlier, there are many sectors that could benefit from this release of surplus labour: for example, the booming geriatric population which could benefit immensely from more healthcare workers and the hospitality industry which could expand massively to match the enhancement of leisure facilitated by computerization. But this calls for both careful planning and policy: labour markets do not clear well as supply is often formed on the basis of decisions regarding skill formation taken in the distant past whereas demand is determined by factors which are more contemporary in nature. I shall come back to this discussion later.

Identifying and measuring the impact of computers and robots on economic growth

While Spence's arguments suggested a significant contribution of innovations in AI – hardware, software, and networks – to the growth rates of per capita GDP in various economies, empirical data did not point to much of an elevation in such growth rates; there was some evidence of an AI-induced boost in productivity in the 1980s and very short-lived high growth in productivity in 1995–1999 (Lowrey, 2011).

Why did per capita GDP growth not register a persistent increase even though the AI revolution was characterized by a vast increase in the number of transactions between providers of products and services and consumers? The reason is actually quite simple: GDP by convention has been calculated on the basis of revenue generated through sales of final goods and services, the assumption being that such revenue was an indicator of value generated. But after the internet became available to all, music, movies, documentaries, blogs, vlogs, and social media accounts – all satisfying ways of spending one's leisure time – either became available for free or at a very low price. While consumers were now spending a very negligible proportion of their income on such products, as compared to significant amounts spent earlier on music and video CDs or social interaction often involving travel, they were actually consuming much more of videos and music and interacting way more with the outside world than before. The changes in per capita GDP thus did not satisfactorily capture the boost to standard of living imparted by the AI revolution: it was difficult to capture much of the AI-induced increase in consumption of services, old or new, through revenue streams (Lowrey, 2011). It, however, needs to be mentioned that consumption of music, videos, and the like before the internet age was based on the use of material inputs for production of tangible outputs such as CDs and radio sets, and therefore bred forward and backward linkages. As the internet age dawned, the importance of these linkages that generated incomes in the rest of the economy waned; as information was digitized the only significant ones that remained gave impetus to the production of computers and mobile phones. The computer/mobile phone/smart television has emerged as a single window for consumption of most leisure based on sound and images: apart from one of these durables, what is required is only a source of electricity and an internet connection. With the passage of time the proportion of consumed services provided solely through digital means should increase and consequently we should see the economic values of linkages declining and neutralizing some of the positive economic impact generated from free provision of an increasing volume of services.

One way of arriving at the true dollar value of products now supplied free or via low priced subscriptions through the internet is to find out the consumer surplus associated with their use: how much consumers would be willing to pay (WTP) for these products over and above the nominal amounts actually paid if access to any of these through the internet was not open and therefore free. There is a problem here: consumers have become used to not paying or paying nominal amounts for these services and might therefore state WTPs which grossly understate their satisfaction: it is extremely difficult for people to simulate a world inside their brains in which they have to pay significant amounts for consuming the mentioned products and use this simulation to come up with WTP measures for such a hypothetical world which would be sensitive to prices of other products and their own levels of income.

A study by McKinsey Global Institute (2011) came up with a survey-based estimate of the value generated by a large set of services provided free through the internet: email accounts, provision for browsing and search, blogs, and social media. Even though these measures are bound to be underestimates because of the mentioned reasons, their magnitudes are significant: in Germany, France, the United States, and the United Kingdom the per user values were 13, 16, 19, and 20 USD per month, respectively, and the corresponding annual aggregates for the mentioned national economies were 7, 7, 46, and 9 billion USD, respectively.

The same study was also able to use the cross-sectional variation in growth of per capita GDP among 13 of the world's major economies – the developed countries covered being the United States, Canada, Germany, France, the United Kingdom, Italy, Sweden, Japan, and South Korea and the fast-growing developing economies covered being the BRIC economies of Brazil, Russia, India, and China – to show that these growth rates were positively correlated with a measure of the depth of internet access defined as a weighted average of (i) an index of the quality of infrastructure available for accessing the internet and (ii) the intensity of its usage by individuals, businesses, and governments. A regression analysis, controlling for growth of fixed capital per capita and labour input per capita in the nine developed countries studied, revealed that the impact of the mentioned measure on the growth rate of per capita income was not only positive and statistically significant but also of a large magnitude. The study also explored the micro foundations of these results through a survey of 4,800 small and medium-sized enterprises (SMEs) in 12 out of the 13 mentioned countries. The results revealed that SMEs which used web technologies grew significantly faster than others.

A recent discussion paper by McKinsey Global Institute (2021) also points out another key reason why economic growth, as measured conventionally through growth of GDP, has slowed down: most companies and sectors devote a lot of time to generate intangibles such as blueprints and software which are not valued in the short run by national income estimates. However, these form part of plans which should fructify in the distant future. In other words, the growth dividends of the digital revolution can be reaped in the distant future. But whether this actually manifests itself in the growth statistics cannot be predicted as of now because as mentioned, the number of free goods has increased and might keep on increasing.

Effects of robotization on equilibrium wages and employment

A meta-analysis reveals that a very significant proportion of workers are at risk of losing their jobs to robots in the next 20 years in developed countries:

for the United States, Frey and Osborne (2013) calculated this proportion to be 47% while a study by McKinsey claims the same statistic to be 45%; for the Organisation for Economic Co-operation and Development (OECD) as a whole the World Bank estimates this proportion to be 57%. These statistics just tell us what proportion of human jobs is replaceable by robots, given possible technological progress in regard to robots. The studies mentioned in this paragraph need to be supplemented by studies which look at (i) how wages will move, and prices of robots change over time and (ii) how productivity of human labour will change with the passage of time, with both (i) and (ii) dependent on the nature and extent of automation and determining how the mentioned risk gets actually translated into a net loss of jobs.

One such study which looks at how robotization changes the wage-employment outcomes in labour markets is by Acemoglu and Restrepo (2017a, b). They use the IFR's (International Federation of Robotics) definition of robot: "an automatically controlled, reprogrammable, and multipurpose [machine]". In their study they only consider industrial robots which undertake the jobs previously performed by manual labour through limb movements: welding, painting, assembling, handling materials, or packaging. This leaves out hardware and software capable of displacing humans from jobs associated with the performance of cerebral work. Given that such software and associated hardware can be considered as part of AI, the impact of robotization on the labour market has been underestimated by the authors.

The authors identify various important effects of robotization: by displacing humans from employment in certain sectors, robotization enhances the mass of human population seeking jobs in non-robotized sectors, leading to a dip in the wage rate and providing incentives for increasing human employment and output in these sectors. Robots also enable employed humans in the economy to be more productive in their jobs: thus, robots not only displace humans from their jobs but also make humans who are not displaced from their jobs or added to the workforce more productive. This increase in productivity can provide the stimulus for job creation that can offset some of the job destruction caused directly by the introduction of robots.

Their study is used to gauge the impact of these effects by looking at how variation of exposure to robotization across local labour markets in the United States is associated with that in outcomes in regard to wage rate and employment. The exposure of each local labour market to robotization is calculated as a weighted average of penetration of robotization in national industries, the weights being the employment shares of these industries in the local labour market. Consider, for example, an industry A which is characterized by highly robotized production and another industry B characterized by the absence of robots. Also consider the following stylized example: local labour market 1 is only different from local labour market 2 in the sense that the former has a higher employment share in industry A than the latter and the latter obviously has a higher employment share than the former in regard to industry B, with the difference in employment shares of the two markets in regard to industry A exactly equal in magnitude to that in regard to industry B. In that case, the measure of exposure would always reveal local labour market 1 to have higher exposure to robotization than the other market, an outcome of the larger size of industry A in market 1 and a correspondingly lower size of industry B.

As mentioned, the authors try to determine how the change in exposure to robotization thus measured impacts wage rate and employment by actually computing such exposure on the basis of real data and relating it to data on the wage rate and employment in various local labour markets; it is found that that if a new robot is added to every 1000 workers in a commuting zone (local labour market) then it reduces the local employment-to-population ratio by 0.18–0.34 percentage points and local wages by 0.25%–0.5%. This is equivalent to three workers losing their jobs to every robot.

As mentioned, the discussed study looks at the impact of the introduction of industrial robots in various sectors of the U.S. economy on wage and employment outcomes in the U.S. labour market. Graetz and Michaels (2015) on the other hand conduct a cross-country study: they examine the use of industrial robots in 1993–2007 in 14 industries, from manufacturing to agriculture to utilities, in 17 developed countries including European economies, Australia, South Korea, and the United States.

The authors of this study found that use of robots grew rapidly in the mentioned period as technological change reduced the prices of industrial robots, adjusted for changes in quality, by around 80%. An indicator of robot use is the ratio of the number of robots to hours worked by humans. This indicator increased by 150% during this period. The authors of this study calculated robot density – the stock of robots per million hours worked by humans – for each combination of industry type and country, the unit of analysis. Their econometric analysis upheld the hypothesis that a higher robot density would lead to higher labour productivity and therefore output. It concluded that robot densification (a higher robot density) results in higher total factor productivity – the productivity of aggregate input estimated as an appropriately weighted average of inputs such as capital and labour – and wages and a decrease in output prices.

The authors did not find any impact of robot densification on overall employment in terms of "hours" but there was some negative impact on employment of middle-skilled and low-skilled workers; this was not consistent with Acemoglu and Restrepo's (2017a, b) finding that introduction of robots had a negative impact on employment prospects of all types of workers, irrespective of the skill level they were associated with. Importantly, the authors found that robots constituted only a miniscule portion of the capital stock even in robot using industries; this implied that there was a lot of room for robot densification to occur in response to technological progress bringing about a fall in the price – computed by controlling for changes in quality – of robots.

An important contribution to the literature on impact of robotization on employment and wages has been made by Marin (2014) as her work refutes the conjecture by Acemoglu and Autor (2011) that the recent AI revolution has led to an increase in the wage premium enjoyed by skilled/highly educated workers over low-skilled workers. The reason advanced for an increase in the wage premium is that technology and skilled work are complements. Marin shows, using data from Timmer (2012), that skill premiums in regard to wage have declined in all Western countries, with the exception of the United States and Germany, and have been exhibiting a plateau in the United States since 1999. Marin draws our attention to the fact that in all countries other than the United States and Germany, the supply of skills and education, as revealed by the attainment of academic degrees, advanced at a very fast pace in 1996–2012, outstripping the advance in technology: for example, in Italy and the United Kingdom the share of people with academic degrees in the workforce doubled; and in Spain and France it went up by more than 50%. It is quite obvious that aggressive policy in regard to higher education led to an excess supply of highly educated/skilled people and a downward pressure on skilled wage and therefore the skill premium. But the slow pace of the spread of higher education in the United States not being associated with an increase in wage premium also indicates that there is something significant in the nature of technological progress itself that can bring down the skill premium: as argued by Marin and illustrated in Chapter 2, developments in software and hardware have been such that these can perform at least some of the services provided by human capital – for example, that embodied in journalists, artists, professors, and lawyers. As we have learnt in the previous pages, machine learning can be successfully used to make computer systems substitutes for skilled human workers under a variety of circumstances; when soft skills, useful for dealing with unforeseen contingencies, or feelings such as empathy are required, humans have a vast advantage over robots, and in these cases, robots are human collaborators rather than substitutes. Thus, robotization generates opposing sets of tendencies, one enhancing the skill premium and the other reducing it.

The set of reasons which have led to a decline in the skill premium has also led to an increase in skilled unemployment: in the period 2000–2012, there has been a doubling in the United States and the United Kingdom while in Spain and Italy there has been a tripling. Of special significance is the doubling in the United States in spite of slow advancement of higher education as it shows that the substitution of human capital by AI is significant. No other factor can explain the trimming of the skilled workforce.

The takeaway is that broad and blunt policies associated with the promotion of acquisition of significant skills can be inappropriate; rather the policymaker should survey various types of skills across the board and identify skills that are expected to remain the preserve of humans. It is only in these fields that a push should be given in terms of subsidized skill transfer. I shall discuss this further in the next chapter.

As we have seen, the impact of the AI revolution, also called the Second Machine Age, on the average wage, overall employment and the level of employment by skill type, and the skill premium in regard to wage differ from study to study; the reason for the difference in outcomes is that these studies are characterized by differences in regions studied, time periods covered, and the component of AI considered. It is for this reason that a theoretical framework is needed to look at how AI can substitute for the role of humans in the production process while at the same time increasing their productivity and enabling them to carry out new complex tasks. Acemoglu and Restrepo (2016, 2017c) build an elegant framework in this regard, the subject of discussion in the next few paragraphs.

Acemoglu and Restrepo (2016, 2017c) assume that at any point of time a given set of tasks are carried out in an economy to yield satisfaction for the consumers. These tasks can be arranged in ascending order of human skill content; therefore, as one moves up this hierarchy of jobs, the productivity of the human performing the job increases. The extent of automation depends on two factors: a task is automated if the technology for automation exists, and the cost of performing the task through automation is lower than the cost of getting it performed solely through human capital. As human productivity is higher in tasks involving a higher level of skill, we can say that performance of a task by human capital becomes more cost efficient as the skill level increases; this in turn implies that at any point of time all jobs associated with a lower level of skill than a threshold level will be automated and others will be performed by human capital. This conclusion accords with reality to some extent though not fully: robots in hospitals carrying medicines and documents from place to place while doctors perform surgeries and examine patients is consistent with the conclusion; at the same time, the significant number of lowly paid cleaning ladies or male janitors in the workforce performing jobs involving complex limb movements – bending, pouring, tearing, scraping – that cannot be automated is inconsistent with the conclusion. One can also claim that cleaning ladies and male janitors constitute highly skilled labour on the basis of the vast difference between the complexity of movements that they are capable of and that displayed by robots. However, note that almost all healthy adult humans are capable of complex body movements, such as those performed by cleaning ladies or male janitors; though there is a gulf between humans and robots in the capability to perform such movements there is not much of a variation in this capability within adult humans. Thus, given the highly flexible muscular and skeletal system of humans, the extent of skill transfer, measured in hours and days, needed for a human to become an effective cleaning lady/male janitor is way less than that required to become a doctor, professor, or a lawyer – a few hours or days to become a practitioner in the first set as opposed to several years in the second set. Thus, if we measure the skill involved in an occupation on the basis of the length of the process of required skill transfer, we can unambiguously label the former set as "low-skilled" and the latter set as "highly skilled". This would then violate the Acemoglu-Restrepo conclusion of jobs below/above a threshold in the mentioned hierarchy of jobs being automated/"performed manually".

In spite of these imperfections the Acemoglu–Restrepo (AR) framework remains valuable in analyzing the nature and pace of automation in an economy; note that a satisfactory level of performance in jobs which involve a high level of skill – for example, well performing professors, lawyers, doctors, and policymakers – is often an outcome of the combination of (i) a high stock of knowledge, (ii) well-honed deductive skills, and (iii) ability to express oneself in writing or speech in ways that enable the professional to explain and present the conclusions derived by using (i) and (ii) as well as the ability to perceive and judge human emotions. Machines can neither adequately acquire these skills nor suitably combine them. This can serve as an adequate basis for the AR assumption that the jobs in the occupational structure corresponding to the highest level of skill cannot be automated, at least as of now. At the same time, they are also quite right about robots (machines) picking up some skills whose attainment by humans is based on the expenditure of a considerable amount of time and monetary expenses. For example, consider the development of software which is able to read X-ray plates and thus poses a threat to jobs held by radiologists. Such software tends to narrow the range of skilled jobs that is only open to humans and not to machines.

But importantly, AR points out that automation also opens up a window of opportunity for human jobseekers. They draw our attention to the Second Industrial Revolution which introduced various machines such as trains plying on railroads replacing horse drawn stagecoaches in regard to quick transportation of humans and freight over long distances on land; steamboats replacing the slower sailboats in regard to transportation over water; and cranes, the mechanical alternative to dock workers for lifting freight. But the arrival of these machines led to the emergence of a new set of occupations for maintaining, repairing, and coordinating the output of these new machines: engineers, conductors, managers, and financiers. The same experience of automation giving rise to new occupations/"job titles" is being repeating in recent times: the new jobs include that of programmers, data analysts, audio-visual specialists, computer support specialists, etc. In regard to the U.S. labour market, AR points out that new jobs and new titles account for a large fraction of employment growth.

The empirical analysis by AR for the United States shows that for each decade since 1980 in the period 1980–2007, occupations with more new job titles, which in turn involve the performance of more new tasks than traditional job titles, have displayed greater employment growth: on the basis of regression analysis, they concluded that a 10 percentage point gap at the beginning of each decade in the number of new job titles associated with occupations translated on an average to a 5.05% greater employment growth over the decade. The higher growth of employment in occupations with new job titles as compared to a benchmark category with no new job titles accounted for 50% of the 17.5% employment growth in the U.S. economy in 1980–2007.

The observations about automation and the emergence of new tasks/job titles associated with management and coordination of new machines by humans form the crux of the AR framework: automation taking away jobs from humans but at the same time giving an opportunity to humans to invent new tasks based on the new machines. If the direct effect of automation overwhelms the growth in jobs resulting from the new tasks, unemployment will increase and the share of labour in national income as well as real wages will decrease; if the opposite happens there will be a contraction of unemployment and an increase in the mentioned share of labour as well as real wages of workers.

As AR correctly point out, the race between automation and the creation of new complex tasks is inevitably characterized by certain developments which determine whether the net addition to unemployment caused by the passage of time tends to be zero, positive, or negative. Every wave of automation releases a surplus of labour onto the labour market, bringing down wages and encouraging employment creation through a new set of tasks that can take advantage of the wage decline; once this surplus labour is exhausted and wages go back to their previous higher levels, the economy becomes ripe for another wave of automation and so on. But this tendency for the economy to move in the direction of balance might be neutralized if (a) capital is very cheap and encourages one wave of automation to follow another in rapid succession without interludes in which the mentioned invention of new complex tasks takes place and (b) the technology for coming up with new automation progresses such that automation becomes cheaper and easier to achieve. An example of (b) is the development of the Robot Operating System (ROS), mentioned in Chapter 2, which runs on principles similar to those for operating systems such as Microsoft Windows and provides a very helpful environment for generating and hosting various software for powering robots to do different and new tasks. Add to this the fact that technological progress has resulted in computers and robots being faster, smaller, and portable over time, which in turn has implied that the capital for achieving any task through automation has become cheaper. Thus, while the rich AR framework admits several possibilities, a recent history of automation which showcases the development of operating systems such as Microsoft Windows and ROS points to machine powering ahead of man in the performance of tasks that yield value to man.

Adoption of robots over time and the threat to human employment: what do the numbers tell us?

Using IFR data, Acemoglu and Restrepo (2017a, b) point out that in 2007 the number of industrial robots in use in the United States and Western Europe was four times that in 1993; thus, adoption of such robots grew at a staggering 10.4% in 1993–2007. According to IFR, there were between 1.5 and 1.75 million industrial robots in operation in this region in 2017 while Boston Consulting Group (BCG) estimated that this number could increase to 4–6 million by 2025. We can thus estimate the lower bound for annual growth of employment of industrial robots in 2017–2025 to be 10.88%, which when compared with the figure of 10.4% for 1993–2007 indicates an increase in the growth rate of adoption of industrial robots over time even though the base for determining this growth rate is constantly increasing as a result of positive growth. Given that industrial robots act as perfect substitutes for human labour in certain important sectors, we can consider aggregate labour in use in an economy to be a weighted average of the levels of robotic and human employment. Unless we assume a major increase in the growth rate of the aggregate requirement for such labour the fact that the trend growth rate of employment of industrial robots is around 11%, far greater than the current growth rate of human employment, implies that the growth rate of human employment has to go down in the future. Thus, *ceteris paribus*, on the basis of secular trends in use of industrial robots we should expect the incidence of unemployment in the world to rise significantly as a result of the introduction of industrial robots. The use of artificial intelligence in substituting for white-collar work in a wide variety of service sectors should further exacerbate this trend.

Some number crunching will make things clearer. The identity for aggregate labour employed is given by $A = L + R$ where A is aggregate labour employed, L is human labour employed, and R is the robotic labour employed expressed in terms of equivalent units of human labour. Thus, if r is the number of robots employed, we have $R = ar$, where a is the no. of human labour units that are the equivalent of one robot – for example, $a = 7$ implies that in a given period of time one robot can do the work of seven humans. For the sake of simplicity, let us assume that a remains constant for the time period over which the projection is being made, a valid assumption if the period is short. Thus, we should expect the growth rate of R for North America and Western Europe to be equal to the growth rate of r (10.88%) over the period 2017–2025. Compare this to the global growth rates of human employment for a recent period 2000–2013: the number of employed people grew from 2. 290 to 3.114 billion at an average annual rate of 2.39%, much smaller than the growth rate (10.88%) of the equivalent of man days of work performed by robots for Western Europe and North America, a region covering developed countries which is large enough for this growth rate to be representative of the globe as a whole. With humans still supplying bulk of the labour used in production the growth rate of aggregate labour employed (A), a weighted average of the growth rate of labour supplied by humans and the growth rate of the equivalent of human labour supplied by robots, is far closer to 2.39% than it is to 10.88%. Assuming no changes in the growth rate of A and R over the period 2017–2025, the growth rate of L (human employment) $= A - R$ has to decline as the weight attached to the growth rate of R, higher than the growth rate of A, increases. This obviously implies an increasing incidence of unemployment even in the long run unless measures are taken by governments and allied private sector to generate skills in areas where there already is or likely to be an unmet demand for labour: geriatric care, hospitality, etc.

The *World Bank Jobs Database* reveals that the number of employed people in the United States and Western Europe in 2013 was 147.3 and 173.9 million, respectively, yielding a total of 321.2 million. Assuming growth by 10% in 2013–2017, in keeping with observed data, yields a total employment of 354 million in 2017. The lower bound for the Boston Consulting Group's estimate of increase in the number of robots during the period 2017–2025 is 2.25 million. The AR estimate of a robot displacing three workers implies that in 2017–2025 adoption of robots will reduce growth of human employment in the United States and Western Europe by 6.75 million, approximately equal to 1.9% of human employment in 2017 – significant but not too large. Use of the upper bound of the BC group's estimate implies a number of robots installed of 4.5 million robbing the economies of the United States and Western Europe of as many as 13.5 million jobs – significant as well as large and 3.8% of human employment in 2017. Given the rate of displacement of humans from jobs by robots, as revealed by this discussion, the destruction of human jobs by robots in 2017–2038, a period of 20 years, could well be 10% of total human employment in 2017.

Moreover, the AR estimate of "one industrial robot destroying three jobs" could be considered an underestimate as they do not consider the negative impact of job destruction on aggregate demand and through it the demand for human and machine labour. I shall discuss this issue in a later section. Furthermore, in the future, both the software and hardware associated with industrial robots is surely going to improve, as already discussed. This is surely going to cause an increase in the job displacement potential of an installed robot.

In regard to developing countries, the ILO Bureau for Employers' Activities and the ILO regional office in Asia and the Pacific have together researched the impact of automation and AI on employment in the ASEAN region. Five major sectors have been studied: automotive and auto parts, electrical and electronic parts, business process outsourcing, textile, clothing and footwear, and retail. The study also included a detailed survey of 4,076 responses from ASEAN enterprises in the manufacturing and service industries as well as 2,747 responses from those engaged in university and technical education.

To add to this, there were 50 interviews with key stakeholders in ASEAN countries and validation exercises with executives in Cambodia, Indonesia, and Singapore. This project by the ILO generated the output for a series of papers, the most pertinent being *The Future of Jobs at Risk of Automation in ASEAN* (Chang and Huynh, 2016) which applies research methodology developed by Carl Frey and Michael Osborne of the University of Oxford to analyze the susceptibility of various occupations to automation in the ASEAN region.

While ASEAN is a very tech savvy region with high mobile connectivity, it has not been at the forefront of technological innovation and has used technological adaptation to make economic progress. Some of the more backward countries in the ASEAN specialize in labour-intensive manufacturing and outsourcing work given by the developed world. The growth of labour-intensive industry is threatened by automation and technological progress inside developed countries such as the United States where it is now possible for a single robot to complete the entire process of manufacture of products such as garments from fabric.[3] Thus, insourcing might replace outsourcing in the future, though Marin (2014) shows that recent data indicates no contraction in outsourcing. Note however that this is quite consistent with growth in insourcing in a scenario in which there is economic growth and therefore an increase in the volume of economic activity. Marin gives many examples of insourcing becoming popular in the developed world: Apple shifting back some activity from Foxconn China to Silicon Valley in California; Airtex Design Group shifting part of its textile production from China to the United States; and a survey around 2014 by the management consultancy PricewaterhouseCoopers of 384 firms in the Eurozone revealing that two-thirds had re-shored some activities during the 12 months preceding the survey.

Frey and Osborne (2013) estimated that 47% of jobs in the United States were at high risk from computerization within two decades. The mentioned study uses the same methodology to analyze the case of the five ASEAN countries of Cambodia, Indonesia, the Philippines, Thailand, and Vietnam. The study concludes that 56% of all employment in the ASEAN-5 is at high risk over the next decade or two.

Across ASEAN-5 countries, industries identified as possessing high potential for automation are hotels and restaurants, wholesale and retail trade, and construction and manufacturing. The mentioned study identifies education and training as well as human health and social work to be industries with low automation risk in the ASEAN-5 economies. However, Ford (2015) shows that "human capital displacing" technologies are available for the mass dissemination of high-quality education and evaluation of students enrolled in courses in higher education. This phenomenon might not take place in the short run. It remains to be seen whether educational institutions will adopt such technology, but the Covid-19 pandemic might be a game changer in this regard: universities all across the globe have had to rely on online classes and examinations during this period and the popularity of online learning platforms has increased to make up for the absence of education in physical classrooms (see Li and Lalani, 2020).

Going back to the case of ASEAN, if Western universities offer mass online courses whose completion will lead to the award of degrees and diplomas, students in the ASEAN will enrol in those courses in large numbers. This will have adverse implications for the growth of the traditional education sector in ASEAN and can put many jobs, current and potential, in this sector at risk.

Even within the ASEAN the dominant occupations vary from country to country, and it is inevitably these occupations, which provide employment to a large proportion of the labour force, which are at risk from automation. For example, in Cambodia the occupations which are at the greatest risk from automation are those associated with garment production which dominates the national manufacturing sector and employs close to half a million sewing machine operators. In regard to retail in Thailand, one million shop sales assistants are at risk of losing their jobs; the same is the case for clerical work in Indonesia which employs about 1.7 million office clerks.

Frey and Osborne's basic assumption is that only occupations associated with routine and predictable tasks are at risk from automation/computerization; on the other hand, jobs requiring digital dexterity, creativity, social interaction, and emotional intelligence are not that prone to automation/computerization. This assumption might not remain valid in the long run, say, beyond 2040. However, this book does not deal with the long run.

The mentioned ASEAN study relies on interviews of leaders of industry who might not be aware of developments on the anvil; such awareness is greater among developers of software. Once labour-saving software is available and its services are cheaper than that of human labour, profit-maximizing entrepreneurs will be quick to adopt such software and lay off human labour.

Much more alarming are the developments in China which has the reputation of being the world's largest supplier of factory labour. But it seems that enterprises with production units in China have met with the obvious obstacle of rising wages[4] in trying to ramp up production of various commodities and have endeavoured to get around this obstacle by substituting human labour with robotic labour/AI. This process has been encouraged by fast-paced progress in robotics/AI in the past two decades, with the efficiency of robots and computers increasing at a rapid pace.

China, in spite of still being a developing country, accounts for an extremely large proportion of world's industrial robots actively in use. In 2014, this proportion was 25% with numbers rising at a rapid pace – for example, 2013–2014 saw a 54% increase in the number of industrial robots. This clearly points to job losses for human labour. Some specific examples are as follows: in 2015, *Midea*, which produces home appliances and has its production unit in the heavily industrialized province of Guangdong, planned to replace 6,000 workers with machines by the end of the year; *Foxconn*, a manufacturer of electronic goods for *Apple* and other companies with a robotic factory already operational in Chengdu, declared its plans to raise the automated proportion of its factory work to 70% within three years (Ford, June 10, 2015).

Job destruction in regard to blue-collar work is not being adequately offset by the creation of white-collar opportunities. In mid-2013, Chinese government statistics revealed that half of the country's current crop of college graduates were unemployed. Moreover, official statistics reveal that much of this unemployment is persistent and cannot be explained by the time-consuming nature of job search: for example, in 2013 more than 20% of the previous year's graduates remained unemployed (Ford, January 10, 2015). With the scope for using AI increasing vastly and swiftly there is little hope of job creation in the sphere of white-collar work neutralizing the effects of blue-collar job decimation. In fact, there might be a net decrease in the number of white-collar jobs over time due to the mentioned progress in AI unless drastic changes are catalyzed through government policy in the occupational structure to cater to untapped markets in regard to consumer goods and services.

The trends in employment growth in India do not present a rosy picture. According to Abraham (2017), India's employment growth started slowing down after 2004–2005 as per data collected by the National Sample Survey Office (NSSO). The annual rates of employment growth in the periods 1999/2000 to 2004/2005, 2004/2005 to 2009/2010, and 2009/2010 to 2011/2012 were 2%, 0.7%, and 0.4%, respectively. Due to lack of comparable data for the subsequent period the employment picture can be constructed by piecing together the evidence from the Labour Bureau's Employment and Unemployment Surveys (LB–EUSs), the Quarterly Quick Employment Surveys (QESs), and the revised QESs. The exercise reveals negative growth in the period 2013/2014 to 2015/2016, which can be majorly attributed to significant negative growth in the construction, manufacturing, and information technology/"business process outsourcing" sectors. With growth of GDP not slowing down in this period, there has obviously been a marked substitution of labour by capital. There is enough evidence to suggest that robotization might get mainstreamed in the Indian economy in the near future: computerized automation is not new to India and evidence of its use is available in research done 20 years back (Narain and Yadav, 1997). Thus, the prospects in regard to employment growth do not look rosy and the country might continue to experience negative employment growth in a scenario characterized by significant and positive growth rates of population in the working age group.

Moving to Africa, we see that the onset of industrialization is being facilitated by robotization, as documented by Rodrik (2016) and the World Bank (2016), given that the advantages of using robots over employing humans in regard to several tasks have become apparent to entrepreneurs.

Finally, to get another picture of the overall impact of industrial robots on employment and productivity, I look at the evidence presented by Ghodsi et al. (April 2020) in regard to the influence of such robots on the value chain. Consider a given industry which is related to certain upstream industries: the former produces goods by processing products manufactured by the latter. For example, consider garments and cloth where cloth is produced upstream and then processed to generate garments (downstream). These two industries are therefore located on the same value chain.

Thus, we can think of a value chain, with industries adding value and generating different products as we move down that chain. Using data from the 2016 version of the World Input-Output Database (WIOD) and Socio-Economic Accounts (SEA) as well as that of the International Federation of Robotics, Ghodsi et al. come to the conclusion that an increase in the stock of robots in an upstream industry will generate cheaper output at a larger scale in the downstream industry, thus possibly leading to more value added, profits, and employment in that industry. On the other hand, robotization of the downstream industry will increase productivity in that industry and this will generate more demand for the upstream industry and therefore possibly more value added and greater employment in the industry. Overall, there is every possibility of robotization enhancing employment and productivity in the economy.

What we can take away from this chapter: piecing together theory and evidence

On the basis of the literature discussed in this chapter, I now try to integrate the various contributions into a cogent view on how the AI revolution will affect employment, wages, and economic growth. To do this I need to go back to the Solow–Swan model of economic growth (Barro and Sala-i-martin, 1995). The model assumes that production of output in the economy can be understood by looking at the production of output per worker and then blowing it up by the number of workers. For the time being assume that the number of workers is constant. Thus, capital accumulation at the level of the worker occurs because some constant fraction of the output per worker is saved and the output thus saved manifests itself as investment or gross capital accumulation; as capital accumulation occurs, the nature of the production process is assumed to be such that the increase in output due to each additional unit of capital per worker becomes lower and lower, i.e., as additional units get added to the capital stock of a worker his ability to draw output from each successive unit becomes lower and lower because labour supplied by him is spread over an enhanced amount of capital. As the increase in output drawn from each additional unit of capital becomes smaller so does the increase in the amount of capital accumulation. As a result, the ratio of capital accumulation to the stock of capital – the rate of capital accumulation – goes down.

In reality, population growth and depreciation results in loss of capital per worker. If we assume population growing at a constant rate and depreciation of capital per worker occurring at a constant rate, this tends to draw down the capital per worker at a constant rate. Consider what happens when savings, as mentioned, results in capital accumulation at a positive and falling rate at the level of the worker but this phenomenon is counteracted by depreciation and population growth bringing down capital per worker at a constant rate. This will tend to bring down the net or actual rate of accumulation of capital per worker from initial positive levels to zero. Thus, there is a level of capital per worker where the two opposing forces completely neutralize each other. It is this level of capital per worker which tends to persist over time and thus I say that the steady state net growth rate of capital per worker is 0 and the same is the growth rate of per capita output. Those mathematically inclined can look at the Appendix to this chapter for a mathematical depiction of the Solow–Swan model which in turn is based on the details provided in Barro and Sala-i-Martin (1995).

However, empirical data tells us that the long run growth of per capita income is not zero – for example, the growth rate of per capita income in the United States has been positive over the last two centuries. Therefore, it has been pointed out by economists that technological progress needs to be accounted for since it is technological progress which counters the tendency for successive units of capital to add smaller and smaller amounts to output and drag the growth rate of capital per worker or per capita income towards zero. The technological progress that can be modelled and will result in a steady state growth rate of capital and output per worker[5] is the labour augmenting type: in period t each worker functions as if he is putting in $A(t)$ units of labour with $A(t)$ increasing with t. Assuming that $A(t)$ grows at a constant and exogenously fixed rate x, the steady state rate of growth of capital per worker and per capita output is x.

The growth in $A(t)$ over time occurs because "learning by doing" takes place over time and leads to the creation of knowledge by individual producers which then spills over to others. This knowledge leads to each unit of labour (worker) becoming able to generate a higher level of output from a given level of capital, or equivalently each worker being able to generate more labour while operating a given level of capital. The assumption that the rate of growth of $A(t)$ is fixed at a constant exogenously fixed value, x assumes that "learning by doing" occurs at a constant pace over time; moreover, there is an implicit assumption that the spilling over of knowledge created by "learning by doing" – the transmission of new knowledge from creators to others – also occurs at a constant pace. Here the AI revolution has been a game changer. To start with, we need to first note that "learning-by-doing'" includes research which involves the use of data about the production process and subsequent analysis to generate various types of technological progress associated with both innovation and invention. Because of the introduction of computers and various kinds of software, data processing and analysis has become faster and more convenient. Thus, research tasks, from use of knowledge inputs to generation of knowledge outputs, can be conducted more swiftly. Finally, all research outputs can be transmitted using the internet and much of this transmission is costless; the internet, marked by almost instant transmission of knowledge, represents a huge increase in the pace of transmission.

The process of knowledge creation and therefore technological progress is an unending open-ended cycle of "learning (including research) – generation of knowledge – learning –", with learning in the third stage of this depiction of technological progress combining knowledge generated in the second stage and other elements of the stock of knowledge to generate more knowledge in subsequent stages. As each stage in this cycle takes a shorter time to accomplish because of the AI revolution, technological progress has been speeded up. This implies that the rate of growth of $A(t)$ should have experienced a hike because of the AI revolution.

However, if we look at the rate of growth of per capita income, we do not see any acceleration in the growth of per capita income in the period marked by the AI revolution: use of data from the World Development Indicators (World Bank, 2020) shows that the average annual rate of growth for the U.S. economy was 2.5% in 1960–1989 and 1.5% for 1990–2019. How do we reconcile the slowdown in the rate of growth of real per capita income as we go from the period preceding the AI revolution to the period marked by the AI revolution when the AI revolution is clearly expected to speed up technological progress? The answer lies in what I have already pointed out: the AI revolution has resulted in a lot of utility (satisfaction) enhancing services, such as those providing the use of social media and email accounts and the consumption of music and videos, being made available to the masses without charge; the inconvenience in performing chores such as getting passports and driving licences renewed has been done away with as these can now be completed in quick time over the internet without the need to undertake travel, visit offices, and wait in queues; and the material basis on which the consumption of services and completion of tasks has hitherto rested has been diluted in many cases with meetings often being held online without time-consuming travel and associated consumption of fuels, music and movies being consumed digitally through the internet rather than through purchased material intensive compact discs, and so on. The consequences are thus an increase in leisure time, an increase in satisfying means to spend leisure time which are available free of charge, and a reduction in consumption of fuels and materials which augurs well for the environment and the level of pollution; there are however some minuses as the mentioned technological progress has tended to exert a downward pressure on demand for processed materials and fuels and thus affected the generation of incomes in the sectors involved in processing and extraction. However, the mentioned contraction is quite typical of the evolution of capitalism with new sectors coming into use, some old ones contracting and others booming.

The other threat to economic growth comes from technological unemployment, which is also a problem in itself. Automation in the AI age has not only threatened human employment and adversely impacted wage rates in routine jobs but has also threatened to take over some or all of the tasks performed by humans employed in certain types of highly skilled work, such as those by radiologists, professors, lawyers, and journalists: software can read X-ray plates; beam lectures all over the world, thereby implying that one professor can cater to a large population of students where many professors were previously required; compile case histories for ready reference; and write journalistic articles by referring to relevant data. Thus, automation can lead to a contraction of employment in both jobs that involve a high-skill content and others which involve a lower level of skill. Moreover, as has been pointed out earlier there is a belief that automation can only result in destruction of jobs and decline in wage in occupations requiring a low level of skill; this has led to an aggressive education policy in many countries which has turned out highly skilled humans in large numbers, depressing both wage rates as well as employment. The mentioned unemployment promotes income inequalities and leads to a fall in the standard of living of households impacted directly by job destruction and decline in wage rates. Note also that technological unemployment is a consequence of the rational impulses of firms in diverse sectors to invest in labour-saving capital and reduce costs, but the aggregation of these rational impulses and actions can enhance the proportion of population without the necessary means to earn a living. Of course, one can argue that when a wave of automation releases large amounts of labour from employment it provides a large pool of labour at low wage which can be employed in labour-intensive tasks; such employment arrests the proportion of unemployed labour and the downward slide in the wage rate. However, such stabilization of the labour market is prevented if the technology for coming up with new and cheaper ways of automating production progresses so that diversification of production over time is based on mostly automation rather than a process which balances automation and the use of released labour in human capital-intensive tasks. As I have argued this might be true of recent experience: consider for example, the use of large computers by governments/big business which releases labour from employment; this in turn encourages small businesses to expand or set up shop by employing displaced labour at the lowered wage rates, a process which is interrupted by the invention of laptops and other portable machines that provide cheaper services than displaced humans.

Technological unemployment advancing in significant spurts can adversely impact the aggregate demand facing many sectors in the economy, a paradoxical development given that it is the result of rational reactions of profit-maximizing firms to technological progress. This contraction of demand can of course adversely impact economic growth. In the next chapter, I shall look at policy measures that can prevent this adverse impact – those that can maintain consumption in the face of technological employment so that economic growth is not threatened while facilitating the movement of those rendered unemployed to sectors where their labour might be in demand.

Appendix: the Solow–Swan growth model

I now consider a standard model of economic growth – the Solow–Swan model. In this model the economy's output Y is considered to depend on stocks of capital (K) and labour (L) in the economy. The stock of labour is taken to be identically equal to the size of the adult population. All other things remaining constant, a higher K implies a higher Y and the same can be said about a higher L. Thus, we can write

$$Y = F(K,L) \tag{1}$$

The production process for the entire economy is said to display constant returns to scale – the production process generates the same output from K and L as n production processes acting cumulatively, each using the technology captured by (1) and capital and labour given by K/n and L/n. If we select $n = L$ we can write

$$y = \frac{Y}{L} = F(k,1) \tag{2}$$

where y denotes per capita income and k denotes the capital – labour ratio in the economy.

The rate of change of capital over time is given by

$$\dot{K} = sF(K,L) - \delta K \tag{3}$$

where s is the exogenously fixed propensity to save or the proportion of output that is saved. The rate of gross capital formation over time is given by $sF(K,L)$, assuming that savings gets transformed into investment. Given that depreciation of capital is given by the product of the depreciation rate, δ and K, net capital formation $(\dot{K})$, the difference between gross capital formation and depreciation, is given by the above equation.

Using (2) and (3) we have

$$\frac{\dot{K}}{L} = sF(k,1) - \delta k \tag{4}$$

Further note that the rate of change of capital per worker over time is given by

$$\dot{k} = \frac{\dot{K}}{L} - k\frac{\dot{L}}{L} = \frac{\dot{K}}{L} - nk \tag{5}$$

where $\frac{\dot{L}}{L} = n$ is the rate of growth of labour/population.

Substituting (4) into (5) we have

$$\dot{k} = sF(k,1) - (n+\delta)k \Rightarrow \frac{\dot{k}}{k} = \frac{sF(k,1)}{k} - (n+\delta) \tag{6}$$

Note that $F(k,1)$ represents a production process employing 1 unit of labour and k units of capital where k can be varied. Starting from $k=0$, a small increase in the stock of capital will add a huge amount to output; thus $\frac{sF(k,1)}{k} > (n+\delta)$ and the rate of growth of k will be positive. Thus, k will grow with time, but each successive unit of capital will add less and less to the output $F(k,1)$ and as a result the average product of capital, $\frac{F(k,1)}{k}$, will diminish till $\frac{sF(k,1)}{k}$ reaches $(n+\delta)$. Let k^* be the level of k where this happens, i.e., it corresponds to a steady state rate of growth of k equal to 0 which once attained tends to persist as the value of k remains stuck at k^* and in turn leads to such 0% growth. If capital stock per worker settles at k^* then per capita income settles at $F(k^*,1)$ in the long run, i.e., the steady state growth rate of per capita income is 0.

I now model labour augmenting technological progress as follows:

$$Y = F(K, LA(t)) \tag{7}$$

This implies that in period t each worker is equivalent to $\frac{A(t)}{A(\tau)}$ workers in period τ where $\frac{A(t)}{A(\tau)} > 1$ for $t > \tau$. In other words, each unit of labour in a given period is equivalent to more than one unit in previous periods because of technological progress (the fact that $A(t)$ is increasing in t).

An equation similar to equation (6) can be deduced from (7) which can be written as

$$\dot{k} = sF\big(k, A(t)\big) - (n + \delta)k \Rightarrow \frac{\dot{k}}{k} = \frac{sF\big(k, A(t)\big)}{k} - (n + \delta) \tag{8}$$

The amount of output produced by a single worker is given by $F(k, A(t))$. For each level of k, the amount of output produced per worker increases over time because of an increase in $A(t)$ and as a result we have the downward sloping curve representing $\dfrac{sF\big(k, A(t)\big)}{k}$ as a function of k shifting rightward over time. The amount of capital corresponding to zero growth in capital per worker thus keeps on getting revised upwards with capital accumulation and therefore a steady state of zero growth of capital per worker and per capita income is never reached; every time period thus corresponds to a positive growth rate of k.

Note that $\dfrac{F\big(k, A(t)\big)}{k} = \dfrac{kF\left(1, \dfrac{A(t)}{k}\right)}{k} = F\left(1, \dfrac{A(t)}{k}\right)$ because of the fact that the production process is characterized by constant returns to scale and thus associated with a proportionate change in inputs leading to a change in output by the same proportion. Hence from (8) we have

$$\frac{\dot{k}}{k} = sF\left(1, \frac{A(t)}{k}\right) - (n + \delta) \tag{9}$$

To have a steady state or time invariant rate of growth of capital per worker we need a constant $sF\left(1, \dfrac{A(t)}{k}\right)$, given constant s. This would be the case when $\dfrac{A(t)}{k}$ is constant, i.e., the growth rate of capital per worker in steady state is equal to the growth rate of the technology factor $A(t)$, usually assumed to be exogenously fixed at x.

Note that

$$y = F\big(k, A(t)\big) = kF\left(1, \frac{A(t)}{k}\right)$$

Since $\dfrac{A(t)}{k}$ is constant in the steady state, the steady state growth rate of y is equal to that of k, i.e., it is equal to x.

Notes

1 In fact, European economies have already witnessed a truncation of the work week in the recent past (see Konnikova, 2014).
2 Poverty alleviation in quite a few developing countries, most notably the large ones such as India and China, has been swift in the recent past (Niño-Zarazúa and Addison, 2012).
3 Consider the *Sewbot*, the robot that converts fabric to garment much more efficiently than humans. This robot has been employed by the Chinese clothing manufacturer Tianyuan Garments Company in their plant in Arkansas, United States, which was expected to become operational by the end of 2018. In the plant three to five people will work each of the 21 robotic production lines, as opposed to ten workers on a conventional production line. Each robotic line will also be far more efficient, at 1,142 T-shirts in eight hours as opposed to 669 for a human line. Even with highly paid U.S. workers operating the robotic line the cost of labour per T-shirt will be $0.33 as opposed to $0.22 in Bangladesh. However, given that the plant has been located in the heart of the United States, where per capita demand is very high, a garment produced by the *Sewbot* might well be more competitive in price terms than that produced in Bangladesh as the cost associated with transportation and marketing of the garments produced by the *Sewbot* would be lower. Further, one can expect more efficient sewing robots to be developed in the future, which will mean that outsourcing of garment production by developed to developing countries might dry up considerably. However, these are still early days. For details, see Device Plus (2018) and Zhou and Yuan (2017).
4 China's average wage rose very fast in the middle of the second decade of the 21st century and is now higher than that in Brazil, Argentina, and Mexico (Bulloch, March 3, 2017).
5 The rate of growth, which once attained, tends to persist over time.

6

MANAGING THE AI REVOLUTION THROUGH POLICY

A nuts-and-bolts approach

In previous chapters I have looked at the implications of the AI revolution: robotization and allied developments in regard to the internet as well as computer software and hardware. Very briefly, robots have been introduced in manufacturing to perform the same jobs that humans used to perform in assembly line production, and they are also being seen in the services sector performing tasks that were earlier the preserve of workers such as human cooks in restaurants, shop assistants on the retail floor, and nurses and orderlies in hospitals. Add to this the internet, which has speeded information flows and has thus given a huge boost to the coming together of parties for the completion of mutually beneficial transactions, and shortened cycles of knowledge creation owing to much smaller time lags between the creation of knowledge and its transmission for the creation of even more knowledge through research, a boon for technological progress in the global economy. Note that the speed of research has been enhanced by the faster processing of data enabled by computers whose quality, speed, and capacity keeps on improving with the passage of time.

The last development goes hand in hand with the facilitation of a greater ease of access to higher education and skill formation, especially for those belonging to low-income households and/or located in remote areas lacking a base of human capital. This has been made possible by the opportunity the internet provides to educators to fashion courses, awarding degrees/diplomas to those successfully completing these, based on a series of online lectures and training programmes. With a single resource person, aided by technologies for automation of student evaluation, able to cater to thousands of students scattered all over the globe, an opportunity has been handed to educators to vastly reduce the cost of education. This cost reduction coupled with competition among educators to capture the market will ultimately be manifested in a dip in fees that students have to pay for enhancing their skills and knowledge and earning degrees and diplomas.

DOI: 10.4324/9780429340611-6

A huge push will thus be given to the rate of human capital formation and the bridging of interregional differences in the stock of human capital per capita. As human beings, on an average, become smarter over time the development will be paralleled by computers becoming smarter, i.e., faster and smaller because of cutting edge developments in software and hardware, a tendency referred to as Moore's law. The smartening of humans and the machines they handle at hitherto unprecedented pace augurs well for the pace at which labour productivity improves over time.

Much of this progress − the development of the internet and the attendant gains as well as the increasing sophistication of computers and robots − has manifested itself in tangible gains in human welfare, just a forerunner of much more substantial progress in the near future. Already we have seen a large number of new and free products, from email to social media, becoming a part and parcel of our daily lives to the extent that we use them significantly to facilitate work and spend our leisure time. Of vital importance is the development of precision agriculture, which uses AI to accurately calibrate the use of inputs and monitor the needs of plants and soil quality. This promises to contain input costs and bring about a vast improvement in agricultural yields which could be the key to feeding a huge and growing world population and lifting millions, if not billions out of malnutrition and poverty. Vast recent increases in the speed of processing data and the resulting enhanced sophistication of research laboratories also implies that lifesaving research, throwing up cures for hitherto incurable diseases as well as vaccines against dreaded viruses, has been given a boost. Finally, the transition from the "real" to the "virtual" world has lowered the need for travel and exerted a much-needed downward pressure on pollution.

It would not be an exaggeration, therefore, to say that the artificial intelligence revolution has been in many ways a blessing. But it also confronts the human race with significant problems. Policymakers should take note of these problems and prepare to tackle these. Fortunately, the current magnitude of these problems is small enough for the resulting impact to be of limited significance, but complacency and inertia could imply that such impacts can become large and difficult to negate in the future. Fortunately, the evolution to date of the mentioned problems has already yielded valuable data and spurred analysis which can be used for extrapolation of their future magnitudes as well as the formulation of strategies to contain their significance.

The most significant challenge to the human race thrown up by the AI revolution consists of robots taking away jobs from humans in automating (robotizing) sectors and enhancing the residual pool of unemployed workers forced to look for jobs in the rest of the economy. If one looks at a robotizing industry, this might mean that people might not be needed in the jobs for which they have acquired skills: consider what happens when an entrepreneur comes to know that robots can now perform the jobs currently being performed by human employees and at a greatly lowered cost per unit of output without any display of idiosyncratic behaviour or protest, even when sudden escalations of the work schedule are introduced to cater to spikes in demand.

Paralyzing this revolution through newly formulated regulatory policies just because of the mentioned problem of technological unemployment would be an extremely foolish endeavour: stagnation is never desirable, especially when the AI revolution is offering major benefits for the human race. The question therefore is whether we can manage the problem of technological unemployment originating from the AI revolution and its further ramifications without sacrificing its benefits. The answer, I argue, is in the affirmative.

It is true that when entrepreneurs hire robots these will replace humans. Thus, opportunities for cost reduction through robotization of the production process in a given sector of the economy would be promptly grabbed by entrepreneurs, and many trained to work for a lifetime in certain jobs might at least temporarily find themselves in the pool of the unemployed looking for a means to earn their livelihood. Most of these job opportunities would exist outside their area of specialization though there could be some cases in which robotization is embraced by large firms and not by small firms, thus implying that some of those laid off by the large firms could find employment in the small firms. But the broad characteristic of technological unemployment is of labour becoming redundant in its occupation of specialization and therefore needing to go in for a "second best" employment opportunity elsewhere.

The potential enormity of the problem

When robots, newly discovered software, and faster computers throw people out of jobs and trades that they have spent a lifetime mastering, one cannot but help thinking that there is something unjust about it. But at the same time, one has to accept that this is really part of the churning that takes place in a capitalist society offering dynamism and change instead of stagnation. People at least temporarily losing their livelihoods and then learning to swim in occupations in which they have had no experience is however only part of the cost that humans, societies, and economies will pay for the embrace of artificial intelligence. There is obviously the reduction in the value of labour with entrepreneurs relying more and more on capital – non-human means of production – to keep the wheels of the production process moving, a tendency which would promote greater inequality in the distribution of income. But contemplating these trends with care will surely make us hear other alarm bells. When a person gets relieved of a job it takes her sometime to land another job – something which is known in an economist's parlance as search unemployment. But the gravity of the problem is enhanced when the displacement of people from their jobs happens because of technological reasons – say, robots being able to perform the job at a lower cost and not posing the challenges that man-management generates. With no jobs to be found in the industry or sector for which one has been skilled, the duration of the search is enhanced: there is a need for reskilling and acquisition of knowledge before one is ready to take up another job or in many cases start a business. With no help from the government many of those rendered unemployed can fall by the wayside. As the pool of unemployed people swells, industries not majorly affected in contemporary times by the AI revolution can take advantage of the lowering of the wage resulting from job displacement and hire cheap labour. The period devoted to searching for jobs and reskilling can obviously be quite long and those involved would be characterized by low purchasing power and poor quality of life for long periods in the absence of government intervention. Further, the resulting adverse implications for aggregate demand could pose challenges for macroeconomic health.

Note that any unemployment of a significant magnitude and lasting duration can significantly limit the magnitude of aggregate demand in the economy, the simple reason being that those displaced from jobs now have more limited means or incomes to buy things. But this is not a one-way street. One should realize that the growth of the reserve army looking for jobs is a boon for the sectors not experiencing technological dynamism – it gives them a chance to hire more people at reduced wages and produce more. In short, outputs tend to increase while prices tend to decrease. Things are however not smooth – unemployed workers have to wait and search until they get other jobs; the need to reskill before entering a hitherto alien occupation can not only be a costly proposition for the hirers or the jobseekers but can enhance the duration of search as well as limit the incomes which the new hires can command.

In short, aggregate demand can take a significant beating in spite of the impetus given to sectors not experiencing further automation. And in many cases when a large number of industries embrace new developments in AI at the same time, the incidence of search unemployment can display a huge bulge, with extremely ominous implications for aggregate demand. There are redeeming tendencies though – prices of some products can go down because of the emergence of the mentioned cost-efficient technologies based on robotization or the reduced labour costs facing certain traditional sectors of the economy. These developments tend to raise the magnitude of aggregate demand, but they might be overwhelmed by tendencies that depress demand: the mentioned spike in unemployment of a durable nature and the growing concentration of income in hands that own capital and have a smaller propensity to consume than the rest of the population.

The recession in demand would not amount to much if prices fell at the same rate across all sectors in the economy keeping relative prices unchanged as this will not make much of a difference to aggregate purchasing power. But that is unlikely to be the case as unemployment is bound to hit demand for items of mass consumption more than others. Here I am talking about goods and services consumed by the middle, lower middle, and poor classes who would bear the brunt of the burden of technological unemployment – among such Indian families you could witness food consumption restricted to just rice and native bread from a much more varied diet consisting also of fruits, pulses, eggs, etc. and the purchases of consumer durables such as bicycles and two wheelers taking a nose-dive; in American families you would see purchases of new cars, air conditioners, and washing machines deferred. With entrepreneurs producing or selling these items of mass consumption experiencing losses because of reduced demand, it could be curtains for many of them in the long run. With enterprises closing down and labour being laid off, aggregate demand could take a further hit. The economy could possibly get caught in a recessionary spiral.

Of course, this is only one of the possibilities. The factors tending to boost aggregate demand could well triumph over those impacting demand through the generation of unemployment. But the governments of the world have a major role to play in avoiding a doomsday scenario by looking out for significant adverse developments in regard to aggregate demand and employment and coming together to enact policies that minimize the pain felt by people in an interconnected world. I talk about these policies in the next section.

The solution

When workers get displaced from their jobs by entrepreneurs embracing AI it might take a long time before they find jobs again. As mentioned, displaced workers have to find jobs they have not been trained for or even set up businesses they have not been prepared to handle – for instance, software developers rendered redundant by machine learning, which enables the robots to program themselves to a large extent, might have to think about starting businesses in the booming hospitality industry. If a former software developer wants to set up a spa because he perceives a lot of demand for the services traditionally rendered by a spa, he has to look for credit and a suitable location and supervise some minor or major construction so that a spa-type environment could be provided to clients besides recruiting skilled masseuse. Going from being a salaried employee engaged in a job that one has been trained to perform from one's early years in life to setting up a hitherto alien business is a demanding and time-consuming proposition. Imagine what would happen if the same person, after being displaced from the job of a software developer, was picked up by a company and nurtured to be a salesperson. Here the interval of time lying between displacement from one's original job and recruitment in the mentioned new job might not be unbearably long but the struggle daunting: there would be a training period which would not only feel like starting life all over again but also be characterized by a low salary, given that salaries paid by entrepreneurs are ceteris paribus, roughly in proportion to the relevant human capital resident in an employee.

What does all this mean? The upshot of the above discussion is that person displaced from their jobs due to AI might have their lives turned upside down and experience prolonged dips in purchasing power. When this happens, aggregate demand takes a beating and this adverse impact is not, as mentioned, felt uniformly over all sectors. With entrepreneurs able to make more profits initially due to the adoption of AI technologies that economize on cost, the demand for fancy cars, luxury yachts, dream vacations, and expensive jewellery originating from owners of capital might not decline. But at the same time, as mentioned, the demand from those who earn their livelihood solely by flexing mind and muscle to generate labour services might take a beating. And this would be reflected in the sales of television sets, two wheelers, mobile phones, and staple foods: with budget constraints becoming tighter and people being forced to dig into their savings to keep afloat in life, new purchases, especially those undertaken to replace old and depreciated durables, might be put off.

With demand receding in sectors that produce items for mass consumption and possibly overwhelming the positive tendencies generated by the embrace of AI – the fall in costs and therefore prices in robotizing sectors and the reduction in labour costs for non-robotizing sectors – entrepreneurs in sectors producing these items might see a fall in their profits. It is possible that many of them might start making losses as a result of the fall in prices, thus forcing them to exit the industry. When that happens, more employees would be laid aside, resulting in a further reduction in aggregate demand which could in turn further enhance the problem of unemployment and so on.

Whether something this tragic would happen of course depends on a large number of factors. First, whether or not there is a clustering in adoption of AI technologies is crucial; such a clustering would enhance the strength of recessionary tendencies so that these can overwhelm positive tendencies arising from cost reduction induced by robotization. Of course, governments can do a lot in averting such a tragedy: see Box 6.1 for expert opinions in this regard. A vector of tax rates on the use of robots could be put in place in different sectors of the economy and fine-tuned over time so that large bulges in unemployment are avoided. There has to be a dynamism in regard to these tax rates as it would be retrogressive not to embrace technological dynamism and its implications for the march of human civilization; the idea would be to ultimately allow each sector benefitting from the new technology to be able to access it while ensuring that the number of parties adopting the AI technology at any point of time is not excessive, given that excesses would get translated into bulges in unemployment and significant recessions in demand.

There is another reason why dynamism in the vector of tax rates, each one for a different sector potentially benefitting from the AI technology, is probably the best way to go about things: not only is a major recession in demand originating from the generation of unemployment prevented, the basic idea is that these tax rates would all ultimately go to zero. The use of this dynamism would put some entrepreneurs in the waiting list for adoption of the AI technology so that there is no explosion of unemployment with all its deleterious consequences; but at the same time, human capital formation geared to take advantage of the numerous opportunities thrown up by adoption of the new AI technologies would not be deterred. Thus, if we peer through the looking glass, we might see the future generation employed in an array of occupations not only engaging the human and robot in harmony but based on a full exploitation of the complementarity between man and machine, a far cry from the initial scars suffered by the previous generation of humans due to the adoption of robots. There is nothing new about this: we saw in the previous chapter that men turning out more sophisticated machines has been historically accompanied by the emergence of new job titles, with humans engaged in these new occupations able to shepherd the combined efforts of these machines to cater to human wants that had hitherto not been satisfied.

BOX 6.1: TACKLING TECHNOLOGICAL EMPLOYMENT AND ITS CONSEQUENCES – EXPERT OPINIONS

Bill Gates has advocated a tax on the use of robots by employers so that a relative disadvantage of hiring humans, as compared to robots, for the production process is converted into a relative advantage in certain economic sectors, thereby curtailing the magnitude of unemployment generated. He has also suggested the use of proceeds from the robot tax to employ humans.

Elon Musk predicts that robotization can lead to mass unemployment and therefore suggests the use of a universal income scheme to alleviate the resulting economic and social stability.

Lukas Schlogl and Andy Sumner of King's College London, in their study for the Center for Global Development, point to the huge loss caused to developing countries by robotization: robots in developed countries displacing cheap labour in developing countries producing goods for developed country populations (Condliffe, 2018). They suggest the introduction of a universal basic income scheme to protect those rendered unemployed by robotization. This scheme would involve a net positive transfer of income from developed to developing countries.

Guerreiro et al.'s (2020) recommendations provide a valuable input for the framing of suggestions in this chapter: in the short run, the occupations that people are fit for in a robotizing world would depend on (i) choices regarding skill acquisitions that have been made in the past without any knowledge of how automation would affect the appropriateness of those choices in the present, and (ii) taxes imposed on robots to shrink or reverse the advantage of hiring robots instead of humans. Given that it is unfair for employees to suffer because of choices made in the past without any means for anticipation of automation, these employees, according to the authors, should be protected by imposition of a short run tax on robots. At the same time, these authors are against the imposition of a long run tax which would discourage skill acquisition needed to equip humans with the wherewithal for deriving significant benefits from the AI technology.

Some of the suggestions given above were also made by Mitra and Das (2018). In this book I come up with a comprehensive set of suggestions through perusal and synthesis of the contributions made by the mentioned experts as well as Mitra and Das (2018).

However, robot taxes, when fine-tuned, do not just prevent major bulges in search unemployment and the consequent serious recessions in aggregate demand but also generate revenue for boosting consumption expenditures of those rendered unemployed: this evens out the relative fortunes of those remaining entrenched in occupations not witnessing further automation and others who are at the receiving end of brutal job-displacing technological change. Needless to say, such use of revenues prevents the recession of demand originating from technological unemployment, especially that in sectors producing goods and services for mass consumption. As receding demand can result in losses for enterprises and their exit from industries in the long run, the boost to consumption given by robot taxes can make a major difference to the trajectories followed by the economic fortunes of nations. Some of the tax revenue collections can also go towards equipping those displaced from their jobs with suitable human capital so that they can find suitable employment in sectors characterized by unfulfilled demand – for example, hospitality and geriatric care. In short, robot taxes can check impoverishment, surges in income inequality and major slowing down of economic growth.

It might be a good idea to equip many of those displaced to become entrepreneurs. Training people to become employees again in new sectors might not be a good idea; the success of such an endeavour would depend on whether there are enough entrepreneurs willing to set up and expand businesses in these new sectors, even those characterized by a large potential demand. When the government is involved in programmes which provides many of those currently unemployed with the wherewithal to become entrepreneurs it solves the problem of generating enterprise as well as non-entrepreneurial human capital, both important for the expansion of economic sectors, at the same time. Note that the AI revolution marks the enhancement of uncertainty associated with flows of income from human capital which makes it advisable for people to diversify their capital stock to include some physical as well as financial capital. Government facilitated rehabilitation of those rendered unemployed by AI should try to aid such diversification through suitable expenditure of revenues from robot taxes.

The tentative blueprint for government action, thus constructed, can be used to contain recessions of demand while offering much-needed relief for workers displaced by the AI technology by bolstering their consumption levels; wise use of revenue collected through the robot tax can also help to generate personnel and entrepreneurs in sectors where somehow supply has not matched demand, a few examples being the care of the geriatric age group and certain sections of the entertainment and hospitality industry which can cash in on the vast swathes of time liberated by the emergence of time-saving technologies.

Fine-tuning the solution

If the economy being analyzed was the only economy in the world, then the proposed solution can be expected to work well. But if one looks at the national economy to be part of an interconnected whole, it is obvious that the solution would only have the appearance of being promising but would fail when put into practice. Any country levying a robot tax in a given sector will inevitably become a victim of entrepreneurial flight in these times of capital mobility. This is a major reason why the guardians of national economies might be very reluctant to impose the said taxes.

But these taxes do need to be imposed for the myriad reasons I have discussed. There is no alternative to working out sector-specific robot taxes through global agreements, in which case the equality of robot taxes all over the world would prevent the mentioned capital flight. This also makes sense because of the contagion effects let loose by the introduction of robotic technologies – for example, imagine what would happen if major developed countries in North America and Europe robotized production of garments rendering scores of workers in the garment industry in Bangladesh and India jobless.

In short, what one is talking about is a *Robotic Change Fund*, much like *The Green Climate Fund*, with various countries spread all over the world contributing to it. Many aspects of climate change and the embrace of robotic technology are such that we can characterize these as global bads; there are however significant pros in the case of robotization necessitating an attempt to manage rather than tackle it.

If a *Robotic Change Fund* financed by robot tax revenues is indeed set up, a contentious issue could be the way in the which the fund would be augmented and spent. Investing a significant portion of the fund in reliable projects and investments to augment it seems like a sensible idea, given the many useful ways in which the fund could be spent. Spending the fund should be tailored to neutralize the damage from robotization: this global fund would be best used to ameliorate the pain of those rendered unemployed. This in turn implies the allocation of funds in every time period across regions in proportion to the magnitude of technological unemployment caused by the embrace of robotic technologies normalized by the costs and the standards of living.

For the sake of simplicity, consider a world consisting of two countries, say Germany and Bangladesh. Thus, for example, if many more people are rendered unemployed in Bangladesh than Germany by the adoption of technologies that use AI rather than human intelligence, not just inside these economies but all over the world, then other things remaining constant, there should be a greater allocation of this fund to Bangladesh than Germany. But if we keep in mind the reality of German wages being much higher than Bangladeshi wages, it would make sense to multiply the number of jobs lost in Germany by a factor equalling the ratio of the German wage rate and the Bangladeshi wage rate to arrive at the magnitude of German unemployment in terms of the number of Bangladeshi job-equivalents. This can then be added to the number of jobs lost in Bangladesh to get the combined loss in Germany and Bangladesh in terms of Bangladeshi job-equivalents, the shares of each nation in this total worked out and disbursals of robot tax revenue made to each nation according to these shares.

It is worth emphasizing again the different ways in which the fund needs to be spent to get the maximum bang for the buck: the allocation of basic incomes to the unemployed is necessary to help them tide over the period of joblessness as well as maintain the level of aggregate demand in the economy; helping people acquire human capital to become gainfully employed in new arenas and replace the human capital made obsolete by AI is another useful way to spend funds; there is also a dire need for the government to make sure that people become owners of physical and financial capital, given that the mere ownership of wage labour is a risky way to earn a living over long periods of time in the age of AI. Thus, when the government transfers incomes to those rendered unemployed, part of the dole should be in the form of claims on capital, mutual funds, and bonds. While traditionally we have looked upon ownership of financial capital as risky and not something the middle class and the poor indulge in, the time has come for a change in this stance. Large tax breaks for investing in capital also seem the way to go.

Conclusion

When robots displace humans from jobs acquired through considerable training and skill formation, it represents a loss for the economy. Yet this does not provide adequate justification for stalling the further adoption of AI, given the benefits we can reap from AI in terms of a decline in resource costs associated with production, enhanced quality of service delivery, and major potential breakthroughs in regard to health and nutritional outcomes.

Unemployment is a social evil but when it arises from a switch in technologies by profit seeking entrepreneurs it can also be ominous for economic development. Major bulges in unemployment can arise when several sectors in the economy switch to the AI technology at the same time. Once these bulges occur, they can lead to a significant and persistent decline in demand and affect economic growth and development.

However, the government can avoid these bulges and smoothen out a potentially lumpy process of robotization — an array of tax rates that change over time can be used to generate a queue for adoption of AI, and the array can be suitably tweaked over time so that cost-efficient AI technologies can be ultimately adopted. This amounts to a calibrated embrace of AI with the mentioned bulges not occurring. Thus, the recession of aggregate demand can be checked to a significant extent; the revenues generated from robot taxes can act in the same direction as these and can be used to prop up consumption expenditure of the unemployed and enable them to acquire skills for new jobs not likely to be impacted by robotization in the near future as well as become owners of financial capital. Such a trend in ownership of financial capital translates to ownership over a growing mass of physical capital becoming progressively decentralized over time. Such decentralization is a necessity when more and more firms are induced by the profit motive to switch from human labour to its close substitute, physical capital in the form of robots.

Robot taxes cannot be implemented by national governments unilaterally – any inequality in robot taxes will lead to capital flight to places where taxes are lower. The possibility of capital flight will induce the countries of the world to compete in regard to the magnitude of taxes, the outcome of this competition being the plummeting of robot taxes to zero. It is therefore required for the governments of the world to come together and administer these tax rates to ensure that these are the same in any given sector in every country of the world, generate adequate revenues, and vary in regard to their sector-wise magnitudes as well as over time to smoothen out adoption of AI technologies.

7

THE IMMEDIATE FUTURE IN THE LIGHT OF COVID AND THE BOTTOM LINE

Quite often exogenous and unanticipated events have implications for adoption of new technologies. Consider the impact of the 'global recession' in the first decade of the 21st century during which the net number of jobs created in the entire U.S. economy was practically zero. If the recession had not happened, entrepreneurs would not have terminated the jobs on the scale they actually did as mass laying off of workers is viewed as bad for the morale of those already employed, with adverse implications for productivity. Thus, the embrace of available cost-efficient labour-saving technology would have been half-hearted.

However, when demand registered a steep fall during the recession, employers had no option but to terminate a significant mass of jobs. When the economy emerged out of the recession there was no pressure on the employers to go back to the pre-recession level of employment while expanding output; clearly, a switch to lower labour intensity could be achieved by employing more machines and robots without significantly laying off workers and demoralizing the work force. The employment of more robots was deemed more desirable than creating new jobs as robots were easier to manage than humans and could work faster and without breaks. Thus, the Global Recession can be seen to be an unanticipated phase in human history which marked a quantum jump in the employment of robots.

In March 2020, just as I thought I was working on a tight plan to write this book, the Covid-19 (Coronavirus Disease, 2019) pandemic happened. This too was almost completely unanticipated. But I could, after due contemplation, see that this pandemic too had clear long-term implications for human employment and robotization. Thus, a chapter on the implications of Covid-19 for robotization was included.

Before I go ahead and analyze the implications of Covid-19 for technological adoption and innovation, it would be pertinent to outline the basic features of

DOI: 10.4324/9780429340611-7

the pandemic and its impact on morbidity and mortality (see Kumar et al., 2020). The first trace of the coronavirus in the human body was discovered in 1960 and it was seen as causing a cold. Until 2002, coronavirus was treated as a simple non-fatal human virus. In 2003, 1,000 cases of coronavirus and its manifestations in the form of *severe acute respiratory syndrome* (SARS) were reported from the United States, Hong Kong, Singapore, Thailand, Vietnam, and Taiwan. Covid-19, the strain of the virus, which is the cause of the pandemic in 2020, was first identified and isolated in pneumonia patients belonging to Wuhan, China.

According to Chinese Government data, the first case of Covid-19 evidently occurred on November 17, 2019. Until June 8, 2020, more than 7 million cases had been reported across 188 countries and territories and 5.77% of these cases, slightly in excess of 0.4 million, had resulted in deaths (Johns Hopkins University, 2020). As of November 9, 2020, the number of cases had gone up to 50.738 million and deaths to 1.262 million (Worldometer: website). This meant that the fatality rate, deaths from Covid-19 as a percentage of the number of cases of Covid-19, had gone down from 5.77% to 2.49%. The discussion of symptoms that follows is based on Grant et al. (2020),[1] CDC, and WHO (2020). The most frequently observed symptoms include fever, cough, and fatigue whereas less common symptoms are aches and pains, nasal congestion, head-ache, conjunctivitis, sore throat, diarrhoea, loss of taste or smell, a rash on skin or discoloration of fingers or toes, and shortness of breath. About 80% of Covid-19 patients recover from the disease without requiring hospitalization whereas the others become seriously ill and develop symptoms which are potentially life threatening: acute respiratory syndrome accompanied by multi-organ failure, septic shocks (organ injury leading to extremely low blood pressure and abnormal functioning of cells), and blood clots. For symptomatic cases, symptoms appear, on an average, about 5 days after exposure, with the time from exposure to onset ranging from 3 to 14 days.

Our discussion of how the infection is transmitted between people and how such transmission can be checked is based on WHO (2020), European Centre for Disease Prevention and Control (2020), and CDC. The virus can spread from one person to another through small, infected droplets from the nose and mouth of one person – produced by coughing, sneezing, and talking – landing in the nose, mouth, or eyes of the other person. After release, the droplets usually fall to ground or onto surfaces rather than travel through air over long distances. Thus, people can get infected with the virus if they touch these droplets deposited on surfaces and then touch their eyes, nose, or mouth. However, though the virus deposited on surfaces can survive for a period ranging from a few hours to a few days, the proportion of the virus surviving dwindles very fast so that it no longer is large enough to cause an infection. The time between exposure to the virus and onset of symptoms is currently estimated to be between 3 and 14 days. Transmission from a person can begin 1–2 days before he/she displays symptoms, though a person is most infectious during the symptomatic period. There are, however, cases of people catching infections and not developing any symptoms.

To what extent such people can transmit infections is a matter which is not well understood as of now.

Checking the spread of Covid-19

WHO (2020) provides an excellent discussion of measures which can enable people to avoid catching the disease or spreading it. Measures include developing habits of maintaining a distance of at least 1 metre with other persons (termed "social distancing" or more appropriately "physical distancing"); not touching any part of the face while or soon after interacting with others; and washing hands with soap or cleansing these with alcohol-based rub. The transmission of this disease is also checked by people wearing masks covering nose and mouth which prevent infected droplets from getting from one human to another. These are especially useful when persons are separated by less than six feet while interacting with each other: droplets can pass from one individual to the other when they are separated by such a short distance unless there is a physical obstruction. Note that the mask is better at protecting others than oneself as the masked person can become infected through one's eyes or by hands coming into contact with infected surfaces and then with nose, eyes, or mouth. A mask needs to be worn when outside the house especially because a person might not know he is infected: remember that symptoms become evident a day or two after infection and it is possible to become an asymptomatic carrier of the infection. Note that it might not be correct for a person afflicted by sensory sensitivities (CDC) to wear masks. The same is true for people engaged in high intensity activities such as running. While CDC lists running as a high intensity activity, common sense tells us that walking while carrying loads, often necessary on warehouse floors, is also a high intensity activity. Thus, we can draw the conclusion that such activities should be carried out only when masks are not worn and therefore only when social distance is maintained between workers carrying loads. However, given the limited area of warehouse floors such social distancing might not be possible. Thus, entrepreneurs have to turn to robotization, as discussed below.

Other measures to check the transmission of Covid-19 include (i) self-isolation by those displaying symptoms of Covid-19 or testing positive or (ii) equivalently quarantining of such individuals so that uninfected persons do not come into contact with them and contract the infection. It is also advised that people should restrict their use of public transport to essential travel and avoid gatherings as the chances of contracting the infection are higher.

Our discussion of how effective a mask is in protecting a human from the coronavirus in the surrounding environment or preventing an infected human from contributing to the viral load in the environment is based on Pappas (2020) and Rettner (2020). Note that if masks are effective in these roles, these would do a lot to check the spread of Covid-19 as there are significant numbers of mobile people who are pre-symptomatic or asymptomatic and therefore capable of spreading the virus into the environment which can then affect those who have hitherto not been infected by the virus (Bai, 2020).

Rettner (2020) points to a new WHO review based on meta-analysis which concludes that social distancing, face masks, and eye protection all seem to check the spread of Covid-19 in the general community. It is important to note that the authors of the review concluded that none of the three practices fully prevent the spread of Covid-19. Nor did the study establish that the three practices could be used in combination to fully check the spread of Covid-19.

Let us now discuss how each of these practices individually reduced the risk of transmission of Covid-19, as pointed out by the WHO review. Without face masks or any other protective practice, the chance of transmission from an infected individual to an uninfected individual interacting with the former was around 17%. Wearing a mask reduced that chance by 14 percentage points to 3% whereas eye protection reduced the chance by 10 percentage points. Social distancing or keeping a distance of more than 3 feet (1 metre) between individuals was a very good means of reducing the chance of infection or transmission; moreover, this chance became lower as the distance was increased in the 1–3 metre range.

Bai (2020) points to a number of experiments facilitated by nature which highlight how wearing masks reduces the spread of Covid-19. The first is a natural experiment highlighted in *Health Affairs* which points to the growth rates for Covid-19 infection before and after mandates for wearing masks in 15 states and the District of Columbia in the United States and brings to our notice that growth rates after mandates were significantly lower than those before mandates. Similarly, one can see that policy and cultural environments in nations which facilitated mask-wearing led to lower death rates from Covid-19. Consider another natural experiment: a masked man with a dry cough, who subsequently tested positive for Covid-19, flew from China to Toronto but did not transmit the virus to any of the 25 people who were closest to him. Finally, consider an incident involving two Covid positive hair stylists in Missouri who served 140 persons. All 142 persons wore masks and none of the customers were infected with the virus.

Note that since the above are natural experiments a significant proportion of masked individuals in the above experiments were probably wearing cloth masks, given the shortages of surgical and N95 masks and difficulty in wearing the latter type of masks. But at the same time some effectiveness of cloth masks in checking the spread of coronavirus is quite consistent with effectiveness which is significantly short of 100%. Our discussion below indeed points to such effectiveness of cloth masks. Moreover, note that viruses can come into contact with the eyes and possibly infect the individual; masks generally do not offer any physical obstruction which separates the eyes of a mask wearing human from the surrounding environment. Finally, note that the above experiments are marked by 100% of the persons with the virus and those that are at danger of receiving it wearing masks; as the proportion of a population wearing masks falls below 100%, the number of people spreading and receiving the virus rises.

Effectiveness of masks in checking the spread of Covid-19

The effectiveness of masks in checking the spread of Covid-19 varies by type of mask, a fact not highlighted by the mentioned WHO review but emphasized by Pappas (2020). Out of the three major kinds of masks – N95 masks, surgical masks, and cloth masks which include homemade ones – the first type of masks is more efficient in blocking out viruses than surgical masks which are in turn more effective than fabric masks. Supporting these conclusions are those from studies highlighted by Letzter (2020) which emphasize that the effectiveness of surgical masks in checking the spread of any disease marked by flu-like symptoms is much greater than that of cloth masks. Research also tells us that surgical masks can be quite effective in lowering the amount of viral material released by infected individuals into the environment. A study published in *Nature Medicine* (see Pappas, 2020 for details) was based on asking a sample of individuals to breathe into a conical device for a considerable length of time. The collection on the conical device of tiny particles of diameter less than 0.5 microns called *aerosols* and particles of diameter greater than 0.5 microns called *droplets* was examined for the seasonal coronavirus. The droplet particles and aerosol particles breathed out by 30% and 40% of the sample of individuals revealed the seasonal coronavirus when individuals did not wear any masks; when the same individuals wore surgical masks none of the particles collected from them revealed the seasonal coronavirus. This demonstrated that not wearing a mask resulted in the infected person often releasing the virus into the environment whereas wearing a surgical mask prevented him/her from doing so.

While the mentioned experiment was carried out for the seasonal coronavirus and not for the virus causing Covid-19, it was used by researchers to claim that surgical masks diminish the potency of those infected by Covid-19 in spreading the virus. Thus, it seems that having the entire population or a significant portion of it wear surgical masks might significantly check the spread of the Covid-19 pandemic. However, given that N95 and surgical masks are often in shortage and people other than medical professionals find it difficult to wear the former type of masks, the only type of masks recommended by the CDC for general wearing by the public are cloth masks. Our previous discussion of natural experiments highlights that if viral material is floating around in the surrounding environment, then a cloth mask is a physical barrier of significant efficiency. At the same time, this efficiency is significantly lower than that of surgical and N95 masks, as mentioned above, and therefore also significantly lower than 100%. Because these fabric masks leave a lot to be desired in protecting the masked individual from the coronavirus particles in the air, they are not an alternative to social distancing. Furthermore, humans still have to continue to maintain habits such as washing or sanitizing hands frequently.

The impact of Covid-19 on the workplace

The universal use of cloth masks in a workplace, where 1–3-metre social distancing might not be possible, is not enough to prevent the spread of infection though it might diminish the probability of infection spreading. Social distancing when all workers are humans might not be possible because the need for collaboration during manufacture, service provision, or retailing would almost inevitably involve physically close interactions among these humans. But if we replace a significant number of human workers by robots, then social distancing between employed human workers might be possible. And once the significant cost of investment in robots is borne it would make sense to keep the robots employed once the pandemic is over, given that the operating costs in regard to robots is much lower than that in regard to humans, who have to be paid their wages or salaries, when employed, for every month of employment. In other words, Covid-19 might have a disruptive effect on the workplace unless close human–human interactions are replaced by human–robot interactions; given that this replacement can generate great value by facilitating avoidance of the mentioned disruptive effect, it becomes worthwhile for businesses to invest in robots which would also help to save labour costs in the future. Thus, Covid-19 can mark the switch to an era marked by lower labour intensity of output and a much greater employment of robots per unit of output.

Consider a few cases of robotization spurred on by the Covid pandemic (Thomas, 2020): the use by Walmart, America's biggest retailer, of robots to scrub floors; the use of robots in South Korea to measure temperature and distribute hand sanitizer; the use by fast food chains such as McDonald's of robots as cooks and servers; and the use in warehouses, such as those operated by Amazon and Walmart, of robots for sorting, shipping, and packing. Humanoid robots in warehouses and factories can carry loads with greater safety and more efficiency; moreover, the loads they can carry are heavier than those carried by human beings (Chandrayan, 2020).

Robots can act as an interface between the doctor and patient when the latter carries out diagnosis and treatment; thus, they help maintain a safe distance between the doctor and patient in many cases and reduce transmission of Covid-19 or other infectious diseases (PTI, May 9, 2020). A Bangalore-based start-up, *Invento*, has built a humanoid robot that can engage with visitors to collect medical information, thus doing away with the need for human-to-human interface that can lead to the spreading of the coronavirus, in addition to handing out sanitizers and taking a patient's temperature with a thermal imaging device (Chandrayan, 2020).

Howard and Borenstein (2020) point out that companies such as PayPal are also increasingly relying on chatbots to provide customer service; a similar trend for reliance on robots is being observed for YouTube in regard to content monitoring. Robotic telepresence platforms are being used to provide students an "in-person" college graduation experience in Japan, thus doing away with the possibilities of infection that could have been associated with conventional college graduation ceremonies.

Note that people have had to set aside their worries about robotic technologies and AI as these have been pressed into service on a wider scale in the face of the Covid pandemic as a part of the effort to limit the spread of viral infections (Nitin Srivastava, June 8, 2020). The greater contact with robotic technologies thus facilitated has helped people to overcome their biases against these technologies, a development which augurs well for a wider permanent embrace of these technologies in the post-Covid period as compared to the pre-Covid period.

The impact of Covid-19 on the labour market and the global economy

Our discussion in the previous section concludes that many jobs will be lost due to Covid-19. However, the recent *Future of Jobs 2020 Report* by the *World Economic Forum* (WEF) points out that by 2025 many jobs will be created worldwide so that the adverse impact of the job destruction will be largely neutralized. Between now and 2025 the WEF estimates that 85 million jobs worldwide will be destroyed while 97 million new ones will be created. However, the satisfaction we can draw from the projection of a net creation of 12 million jobs should be tinged with caution. First, this figure hides the possibility of job destruction in many national economies vastly exceeding job creation. This is totally consistent with the global projection by the WEF. Thus, some economies might lose significantly and others might win significantly: consider, for example, the possible Covid-19 induced disruption of labour-intensive manufacturing and therefore the global supply chain providing enough reason for developed countries to switch to reshoring from offshoring, and the high labour intensity of production in developing countries such as Bangladesh and Vietnam possibly giving way to mechanization and robotization in response to the problems of managing a large number of employees packed into small workspaces during the pandemic. Second, in all economies, job destruction and creation might occur at different points of time: adoption of new technologies would lead to a high demand for certain types of skills which might not be possible to satisfy immediately through hiring; the process of displaced workers searching for and getting employed in appropriate jobs might be time consuming, especially when they may need to be trained before they become operational in their new roles.

Experts might feel that the story of employment change, caused by the interaction of the need for displacing workers induced by the pandemic and the opportunities created by technological progress, might be the usual story of a disruption leading to job destruction later exceeded by the magnitude of job creation. But Khanna (2020) points out that the technology embodied in AI has developed to the extent that opportunities for labour saving can be grabbed swiftly and easily: AI algorithms can be easily shared and deployed, and robots quickly installed in factories without any changes in the infrastructure within and outside such factories. This implies that the role of humans in the production process will diminish in many sectors on a permanent basis, with adverse implications for total human employment in the economy.

The WEF report points out that the development of AI has led to a growing demand for skills that include critical thinking, analysis, creativity, originality as well as self-management, and the ability to interact fruitfully with others to solve their problems or facilitate collaboration. Thus, job opportunities are arising not only in AI, data science, and cloud computing but in sales, management, and administration. The data from the report shows that during the pandemic, workers are already transitioning from their jobs into unfamiliar areas. Such transition is obviously dictated by the present circumstances displacing workers in their usual jobs. The government needs to play an active role in this regard both by providing subsidized training facilities for workers looking to enter new jobs as well as social safety nets for them when they are unemployed. This will make transition from an existing job, from which a worker can get displaced almost without warning and therefore time for planning, to a new and unfamiliar one relatively painless, smooth and as quick as possible.

Finally, we need to acknowledge that it is difficult to predict both job creation and reduction. First, as Professors Daron Acemoglu and Pascual Restrepo (see Brown, 2019) explain, it is not true that AI technologies that are job-destroying will also be job-creating: consider for example, the technology that gives rise to self-checkout counters and just displaces labour without creating jobs or enhancing productivity. Second, when AI evolves it might initially be job-creating such as internet and file transfers enabling radiologists to interpret "X-ray plates" remotely. This was a development which provided opportunities to hospitals to expand their business by employing radiologists in remote areas, thereby generating new jobs. But later development of AI, which can now read these plates more accurately than radiologists, has made these remote radiologists redundant. Thus, a significant amount of job creation might lead to jobs which are temporary in nature and might soon outlive their need.

While the Covid pandemic will enhance the application of AI in production processes it is seen that the pandemic has compelled students to become more accepting of online means of education, given the non-availability of offline opportunities. Online learning platforms with a business model that is based on free provision of knowledge and almost complete commercial reliance on revenues from provision of space to advertisers have been given an impetus by the pandemic. The enhanced size of their businesses is going to be maintained in post-Covid times, an era in which the cost of human capital formation would therefore be low and prompt a high rate of human capital formation. This is a welcome development, but human capital formed should be in fields where they cannot be substituted by robots. Government incentives have a role to play in this regard.

A summary

The use of robots in production of goods and services often implies that substantial fixed costs have to be undertaken. The significance of these costs as well as the adverse impact on the morale of the remaining workforce resulting from replacement of a substantial section of this workforce with robots are reasons why the pace of robotization might be slow in ordinary times. The spread of the Covid-19 virus and its potential disruptive impact on the workplace, with workers possibly falling ill in droves if they collaborate closely as usual, does not leave employers with an alternative to transferring key roles in the production process from humans to robots. Once they bear the significant fixed costs associated with investment in robots, it is unlikely that they will replace robots with humans in safer times: robots once invested in will continue to provide labour-saving services, and companies utilizing these will save on labour costs and avoid the difficulties of managing a huge workforce. The need for government policy to provide succour to those rendered unemployed by the Covid pandemic, as discussed above, is essential.

The bottom line

It is now time to summarize what this tiny book has achieved. It has definitely not characterized the increasingly AI-enabled future by making specific predictions regarding growth rates and levels of per capita income, unemployment rates, and incidence of poverty. What it has done, however, is to clearly delineate the varied consequences of adoption of AI, some which would gladden the heart of the welfare economist and others which should set some alarm bells ringing.

Is this enough? Would it not have been better to come up with precise numbers on future levels of poverty, per capita income, and unemployment? The simple answer to these questions is that resting on the pride of capturing the essence of the future in precise numbers reflects a remarkable incognizance of the role of multidimensional policy in shaping the future through its awareness of the present. There is some use of these predictions in the sense that they inform us about the desirability of the state emerging from a business-as-usual scenario, thereby serving as an input that helps us determine the urgency for and the needed extent of policy change. Indeed, this book uses numerical predictions from trustworthy studies in precisely this manner. But it combines these with empirical knowledge about the trends in production technology emerging from the advancement of AI, and the use of economic theory to gauge the impact of these trends on actual practices at the level of the consumer and producer and determine the implications of these impacts, transmitted through intersectoral linkages, for the macroeconomy. There is a clear cognizance that governmental institutions and policy do shape the impacts as much as the mentioned trends in technology.

As this book argues, good policy would be able to tame AI, not only ensuring that the human race continues to benefit from its onward march but by also reducing the significance of channels through which it can enhance unemployment and impoverishment and adversely impact economic growth; on the other hand, bad policy can either discourage development of AI and thus its application for improving the quality of life or fail to adequately regulate its development to contain the mentioned adverse impacts; finally, the absence of policy would give individual entrepreneurs complete freedom to express their rational impulses and as this book argues, the aggregation of such unbridled impulses could stoke the fires of unemployment, impoverishment, and recession. The point being made is that it is not possible to predict the future of the human race over the course of the AI era with any certainty. However, much of this uncertainty is a product of our not being able to predict the actions of our policymakers. Therefore, the most appropriate agenda for the intelligentsia is not just to make predictions regarding the future but to influence and lobby for policy which ensures a more desirable future. This book serves as a starting point in this regard, with an attempt to make this pursuit as practical as possible by using a loose open-ended framework that tries to come as close to capturing the complexity of the real world as is possible.

Thus, what has been achieved is a careful attempt to sketch a policy path that would prevent the world economy from coming off its rails in an increasingly AI-driven world and yet enable a steady progress in welfare levels on the basis of that drive. The policy path presented in the book is based on governments of the world reacting to a changing world through economic incentives that shape technological progress and the fine-tuning of other economic determinants such as the income distribution and skill endowments. I hope that it serves as a starting point for more comprehensive debates on policy in the AI age.

Note

1 CDC is the acronym for Centers for Disease Control and Prevention, United States.

REFERENCE LIST

Abraham, V. (2017). Stagnant employment growth: Last 3 years have been the worst. *Economic and Political Weekly*, LII (38), pp.13–17.

Acemoglu, D. and Autor, D. (2011). Skills, tasks and technology: Implications for employment and earnings. In *Handbook of Labor Economics*. Amsterdam: Elsevier.

Acemoglu, D. and Restrepo, P. (2016). *The Job Race: Machines versus Humans | VOX, CEPR Policy Portal*. [online]. Available at: https://voxeu.org/article/job-race-machines-versus-humans.

Acemoglu, D. and Restrepo, P. (2017a). *Robots and Jobs: Evidence from the US | VOX, CEPR Policy Portal*. [online]. Available at: http://voxeu.org/article/robots-and-jobs-evidence-us.

Acemoglu, D. and Restrepo, P. (2017b). *Robots and Jobs: Evidence from US Labor Markets*. National Bureau of Economic Research (Working Paper No. 23285).

Acemoglu, D. and Restrepo, P. (2017c). *The Race between Machine and Man: Implications of Technology for Growth, Factor Shares and Employment*. National Bureau of Economic Research (Working Paper No. 22252).

Ad Hoc Committee on the Triple Revolution (1964). The triple revolution. *Advertising Age Magazine*. [online] Available at: http://pinguet.free.fr/triplefac.pdf.

African Development Bank (2020). *Democratic Republic of Congo Economic Outlook*. [online]. Available at: https://www.afdb.org/en/countries-central-africa-democratic-republic-congo/democratic-republic-congo-economic-outlook.

Akanegbu, A. and R. Ribeiro (November, 2012) The history of calculators: Evolution of the calculator (timeline). *Edtech Magazine*. Available at: https://edtechmagazine.com/k12/article/2012/11/calculating-firsts-visual-history-calculators

Akhtar, A. (2019). Robots are set to wipe out the most sought-after (and highest-paying) jobs on Wall Street in the next 10 years. [online] *Business Insider*. Available at: https://www.businessinsider.in/robots-are-set-to-wipe-out-the-most-sought-after-and-highest-paying-jobs-on-wall-street-in-the-next-10-years/articleshow/68856869.cms [Accessed 14 May 2021].

Alexander, D. (2018). Robotics will change food production forever, creating a more productive and sustainable model. [online] *Interestingengineering.com*. Available at:

https://interestingengineering.com/9-robots-that-are-invading-the-agriculture-industry.

Autor, D. (2010). *The Polarization of Job Opportunities in the U.S. Labor Market Implications for Employment and Earnings*. [online] The Center for American Progress and The Hamilton Project. Available at: http://economics.mit.edu/files/5554 [Accessed 7 Apr. 2021].

Autor, D.H., Levy, F. and Murnane, R.J. (2003). The skill content of recent technological change: An empirical exploration. *The Quarterly Journal of Economics*, 118(4), pp.1279–1333.

Awal, A. (May 30, 2011). A better class of chauffeur for India's luxury car buyers. *Financial Times*. Available at: https://www.ft.com/content/f4384671-3bff-30ee-ac85-f3899016ca32.

Bai, N. (2020). *Still Confused About Masks? Here's the Science Behind How Face Masks Prevent Coronavirus*. [online] Still Confused About Masks? Here's the Science Behind How Face Masks Prevent Coronavirus | UC San Francisco. Available at: https://www.ucsf.edu/news/2020/06/417906/still-confused-about-masks-heres-science-behind-how-face-masks-prevent.

Bain, M. (2017). Rise of the Sewbots: A new t-shirt sewing robot can make as many shirts per hour as 17 factory workers. [online] *Quartz*. Available at: https://qz.com/1064679/a-new-t-shirt-sewing-robot-can-make-as-many-shirts-per-hour-as-17-factory-workers/.

Barro, R.J. and Sala-i-Martin, X. (1995). *Economic Growth*. New York: McGraw-Hill.

Borenstein, J. and Ayanna, H. (2020). AI, Robots, and Ethics in the Age of COVID-19. [online] *MIT Sloan Management Review*. Available at: https://sloanreview.mit.edu/article/ai-robots-and-ethics-in-the-age-of-covid-19/.

Boston Consulting Group (2015). *The Robotics Revolution: The Next Great Leap in Manufacturing*. Available at: https://www.bcg.com/publications/2015/lean-manufacturing-innovation-robotics-revolution-next-great-leap-manufacturing.

Bowles (2016). Our tech future: the rich own the robots while the poor have "job mortgages." [online] *The Guardian*. Available at: https://www.theguardian.com/culture/2016/mar/12/robots-taking-jobs-future-technology-jerry-kaplan-sxsw [Accessed 14 May 2021].

Brown, S. (2019). The lure of "so-so technology," and how to avoid it. [online] MIT Sloan. Available at: https://mitsloan.mit.edu/ideas-made-to-matter/lure-so-so-technology-and-how-to-avoid-it.

Brynjolfsson, E. and Mitchell, T. (2017). What can machine learning do? Workforce implications. Science, [online] 358(6370), pp.1530–1534. Available at: https://science.sciencemag.org/content/358/6370/1530.full.

Brynjolfsson, E., Hui, X. and Liu, M. (2018). Does machine translation affect international trade? Evidence from a large digital platform. *National Bureau of Economic Research*, Working Paper No. 24917.

Brynjolfsson, E., Mitchell, T. and Rock, D. (2018). What can machines learn, and what does it mean for occupations and the economy? *AEA Papers and Proceedings*, 108, pp.43–47.

Bulloch, D. (2017). China is running out of cheap rural labor and it's because of failed reforms. [online] *Forbes*. Available at: https://www.forbes.com/sites/douglasbulloch/2017/03/03/china-is-running-out-of-cheap-rural-labor-and-its-because-of-failed-reforms/?sh=2758be0b25c6 [Accessed 14 May 2021].

Carr, D. (2009). The Robots are coming! Oh, they're here. [online] *Media Decoder Blog*. Available at: https://mediadecoder.blogs.nytimes.com/2009/10/19/the-robots-are-coming-oh-theyre-here/.

Centers for Disease Control and Prevention (2020a). *Coronavirus Disease 2019 (COVID-19).* [online] Centers for Disease Control and Prevention. Available at: https://www.cdc.gov/coronavirus/2019-ncov/prevent-getting-sick/cloth-face-cover-guidance.html?CDC_AA_refVal=https%3A%2F%2Fwww.cdc.gov%2Fcoronavirus%2F2019-ncov%2Fprevent-getting-sick%2Fcloth-face-cover.html.

Centers for Disease Control and Prevention (2020b). *Coronavirus Disease 2019 (COVID-19): Symptoms.* [online] Centers for Disease Control and Prevention. Available at: https://www.cdc.gov/coronavirus/2019-ncov/symptoms-testing/symptoms.html.

Centers for Disease Control and Prevention (2020c). *Coronavirus Disease 2019 (COVID-19): Transmission.* [online] Centers for Disease Control and Prevention. Available at: https://www.cdc.gov/coronavirus/2019-ncov/prevent-getting-sick/how-covid-spreads.html.

Chafkin, M. (2013). Udacity's Sebastian Thrun, Godfather of free online education, changes course. [online] *Fast Company.* Available at: https://www.fastcompany.com/3021473/udacity-sebastian-thrun-uphill-climb.

Chandrayan, A. (2020). How is COVID-19 impacting and transforming the humanoid robot industry? | Robotics Tomorrow. [online] *roboticstomorrow.com.* Available at: https://www.roboticstomorrow.com/story/2020/06/how-is-covid-19-impacting-and-transforming-the-humanoid-robot-industry/15332/ [Accessed 14 May 2021].

Chang, J. and Hyunh, P. (2016). *THE Future of Jobs at Risk of Automation.* [online]. Available at: https://www.ilo.org/wcmsp5/groups/public/---ed_dialogue/---act_emp/documents/publication/wcms_579554.pdf.

Charniak, E. and Mcdermott, D. (2009). *Introduction to Artificial Intelligence.* Reading, Mass.: Fourth Impression, Pearson Education.

Clements, K.W. and Si, J. (2018). Engel's law, diet diversity, and the quality of food consumption. *American Journal of Agricultural Economics,* [online] 100(1), pp.1–22. Available at: https://academic.oup.com/ajae/article/100/1/1/4209927.

Clifford, S. (2013). U.S. textile plants return, with floors largely empty of people. *The New York Times.* [online] 19 Sep. Available at: https://www.nytimes.com/2013/09/20/business/us-textile-factories-return.html.

Cohen, J. (December 16, 2019), Deep reinforcement learning for self-driving cars: An intro. Available at: https://thinkautonomous.medium.com/deep-reinforcement-learning-for-self-driving-cars-an-intro-4c8c08e6d06b [Accessed October 12, 2021].

Condliffe, J. (2018). Developing Countries may need their own strategies to cope with job-taking Robots. *The New York Times.* [online] 9 Jul. Available at: https://www.nytimes.com/2018/07/09/business/dealbook/automation-developing-world.html [Accessed October 12, 2021].

Cooper, M. (2012). Lost in recession, toll on underemployed and underpaid. *The New York Times.* [online] 19 Jun. Available at: https://www.nytimes.com/2012/06/19/us/many-american-workers-are-underemployed-and-underpaid.html [Accessed October 12, 2021].

Covington, P., J. Adams and E. Sargin (2016). Deep neural networks for youtube recommendations. *YouTube White Paper.* Available at: https://static.googleusercontent.com/media/research.google.com/en//pubs/archive/45530.pdf [Accessed October 10, 2021].

Crafts, N. (1999). Economic growth in the twentieth century. *Oxford Review of Economic Policy,* 15(4), pp.18–34.

Davis, L. (2012). This 450-year-old clockwork monk is fully operational. [online] *io9.* Available at: https://io9.gizmodo.com/5956937/this-450-year-old-clockwork-monk-is-fully-operational.

DeAngelis, S.F. (2014). Artificial intelligence: How algorithms make systems smart. [online] *Wired*. Available at: https://www.wired.com/insights/2014/09/artificial -intelligence-algorithms-2/.

Dormehl, L. (2019). The promise and pitfalls of using robots to care for the elderly. [online] *Digital Trends*. Available at: https://www.digitaltrends.com/cool-tech/robots -caregiving-for-the-elderly.

Edwards, R. (1987). Computer technology and unemployment. *The Australian Quarterly*, 59(1), pp.84–90.

Engel, E. (1857). The production and consumption conditions of the Kingdom of Saxony. Reprinted in *Engel's Die Lebenskosten Belgischer Arbeiter-Familien [Cost of Living in Belgian Working-Class Families]*. Dresden: Forgotten Books, 1895.

European Centre for Disease Prevention and Control (n.d.). *Q & A on COVID-19*. [online] European Centre for Disease Prevention and Control. Available at: https:// www.ecdc.europa.eu/en/covid-19/questions-answers [Accessed 30 Jun. 2020].

Ford, M. (2015). *Rise of the Robots: Technology and the Threat of a Jobless Future*. New York: Basic Books, A Member of The Perseus Books Group.

Frey, C.B. and Osborne, M.A. (2013). The future of employment: How susceptible are jobs to computerisation? *Technological Forecasting and Social Change*, [online] 114(1), pp.254–280. Available at: https://www.oxfordmartin.ox.ac.uk/downloads/academic /The_Future_of_Employment.pdf.

Gamble C. and J. Gao (August 17, 2018), Safety-first AI for autonomous data centre cooling and industrial control. [online]. Available at: https://deepmind.com/blog /article/safety-first-ai-autonomous-data-centre-cooling-and-industrial-control [Accessed October 11, 2021].

Ghodsi, M., Reiter O, Stehrer, R and Stöllinger, R. (n.d.). *Robotisation, Employment and Industrial Growth Intertwined Across Global Value Chains*. Working Paper 177. Vienna Institute for International Economic Studies.

Gonzalez, A. (2018). Worker suicide at Foxconn factory in China adds to concerns about poor working conditions. [online] *GoodElectronics*. Available at: https:// goodelectronics.org/worker-suicide-foxconn-factory-china-adds-concerns-working -conditions/.

Gosset, S. (2019). Farming & agriculture Robots. [online] *Built In*. Available at: https:// builtin.com/robotics/farming-agricultural-robots.

Graetz, G. and Michaels, G. (2015). Robots, productivity, and jobs | VOX, CEPR policy portal. [online] *Voxeu.org*. Available at: https://voxeu.org/article/robots-productivity -and-jobs.

Graham, M. (2019). NBC says it has topped $1 billion in national ad sales for 2020 Summer Olympics. [online] *CNBC*. Available at: https://www.cnbc.com /2019/12/10/nbc-says-it-topped-1b-in-national-ad-sales-for-2020-summer -olympics.html.

Grahame, A. (2017). What countries manufacture Adidas? [online] *Career Trend*. Available at: https://careertrend.com/info-8187202-countries-manufacture-adidas.html.

Grant, M.C., Geoghegan, L., Arbyn, M., Mohammed, Z., McGuinness, L., Clarke, E.L. and Wade, R.G. (2020). The prevalence of symptoms in 24,410 adults infected by the novel coronavirus (SARS-CoV-2; COVID-19): A systematic review and meta-analysis of 148 studies from 9 countries. *PLOS ONE*, 15(6), p.e0234765.

Groom, B. and Powley, T. (2014). Reshoring driven by quality, not costs, say U.K. manufacturers. [online] *Financial Times*. Available at: https://www.ft.com/content /9757ffcc-9fc9-11e3-94f3-00144feab7de.

Guerreiro, J., Rebelo, S. and Teles, P. (2020). Robots should be taxed, for a while. [online] *VoxEU.org*. Available at: https://voxeu.org/article/robots-should-be-taxed -while.

Gupta A. (August 28, 2019). AI learning to land a rocket (lunar lander) | Reinforcement learning. Available at: https://towardsdatascience.com/ai-learning-to-land-a-rocket -reinforcement-learning-84d61f97d055 [Accessed October 7, 2021].

Harris, K. (2021). Vice President Kamala Harris: COVID-19 unemployment calls for biggest jobs investment since World War II. [online] *USA TODAY*. Available at: https://www.usatoday.com/story/opinion/2021/04/12/kamala-harris-covid -unemployment-american-jobs-plan-column/7189777002/.

Heath, N. (2018). What is machine learning? Everything you need to know. [online] *ZDNet*. Available at: https://www.zdnet.com/article/what-is-machine-learning -everything-you-need-to-know/.

Heslin, K (July 30, 2015), A look at data center cooling technologies. [online] *Uptime Institute* (Online) Available at: https://journal.uptimeinstitute.com/a-look-at-data -center-cooling-technologies/ [Accessed October 11, 2011].

Hicks, M. (2018). Microsoft's new AI translates Chinese-to-English as well as a human translator. [online] *TechRadar*. Available at: https://www.techradar.com/news/ microsofts-new-ai-translates-chinese-to-english-as-well-as-a-human-translator [Accessed October 11, 2011].

Hobsbawm, E.J. (1952). The machine breakers. *Past and Present*, 1(1), pp.57–70.

Hockstein, N.G., Gourin, C.G., Faust, R.A. and Terris, D.J. (2007). A history of robots: From science fiction to surgical robotics. *Journal of Robotic Surgery*, 1(2), pp.113–118.

Illumin: A Review of Engineering in Everyday Life (2018). The Da Vinci Robot. [online] *USC Viterbi School of Engineering*. Available at: http://illumin.usc.edu/235/ the-da-vinci-robot/ [Accessed October 11, 2021].

Innovation in Textiles (2017). Automated Sewbot to make 800,000 adidas T-shirts daily. [online] *www.innovationintextiles.com*. Available at: https://www.innovationintextiles .com/automated-sewbot-to-make-800000-adidas-tshirts-daily/ [Accessed October 11, 2021].

Jaimovich, N. and Siu, H.E. (2012). Job polarization and jobless recoveries. *The Review of Economics and Statistics*, 102(1), pp.129–147.

Johns Hopkins University (2020). COVID-19 dashboard by the center for systems science and engineering (CSSE) at Johns Hopkins University (JHU). [online] *Arcgis .com*. Available at: https://gisanddata.maps.arcgis.com/apps/opsdashboard/index .html#/bda7594740fd40299423467b48e9ecf6 [Accessed October 11, 2021].

Joseph Alois Schumpeter (1943). Capitalism, socialism, and democracy. In *The Process of Creative Destruction*. London: Allen & Unwin, pp. 81–86.

Kapoor, Rahul (December 9, 2020), Autonomous driving cars: All six levels autonomous vehicles explained. *Express Drives, Financial Express*. Available at: https://www .financialexpress.com/auto/car-news/autonomous-driving-cars-all-six-levels -autonomous-vehicles-explained-level0-level-1-level-2-level3-level-4-level-5-volvo -chevrolet-audi-bmw-mercedes-benz-artificial-intelligent-ai-self-driving-cars /2146290/ [Accessed October 13, 2021].

Keach, S. (2018). Fast food restaurants are using ROBOT chefs because they can't find enough workers. [online] *The Sun*. Available at: https://www.thesun.co.uk/tech /6625773/burger-robot-fast-food-job-automation/ [Accessed October 11, 2021].

Keynes, J.M. (1930). Economic possibilities for our grandchildren. *Essays in Persuasion*. New York: Palgrave MacMillan.

Khanna, V. (2020). The double whammy of Covid and automation. [online] *The Statesman*. Available at: https://www.thestatesman.com/opinion/double-whammy-covid-automation-1502932627.html [Accessed October 11, 2021].

Konnikova, M. (2014). Why not a three-day week? [online] *The New Yorker*. Available at: https://www.newyorker.com/science/maria-konnikova/three-day-week.

KPMG. (2016). Social Robots. [online] Available at: https://assets.kpmg/content/dam/kpmg/pdf/2016/06/social-robots.pdf.

Kumar, D., Malviya, R. and Sharma, P.K. (2020). Corona virus: A review of COVID-19. *Eurasian Journal of Medicine and Oncology*, 4(1), pp.8–25.

Lee, B.Y. (2016). Current and future trends in vending machines. [online] *Forbes*. Available at: https://www.forbes.com/sites/brucelee/2016/04/22/current-and-future-trends-in-vending-machines/?sh=6039dbcc127a.

Letzter, R. (2020). Can homemade masks protect you from COVID-19? [online] *livescience.com*. Available at: https://www.livescience.com/cloth-masks-coronavirus.html.

Li, C. and Lalani, F. (2020). The COVID-19 pandemic has changed education forever. This is how. [online] *World Economic Forum*. Available at: https://www.weforum.org/agenda/2020/04/coronavirus-education-global-covid19-online-digital-learning/.

Lowrey, A. (2011). The productivity paradox: Why hasn't the Internet helped the American economy grow more? [online] *Slate Magazine*. Available at: http://www.slate.com/articles/business/moneybox/2011/03/freaks_geeks_and_gdp.html.

Mani, G. (2016). Model agricultural land leasing act, 2016: Some observations. *Economic and Political Weekly*, 51(42). Available at: https://www.epw.in/journal/2016/42/web-exclusives/model-agricultural-land-leasing-act-2016-some-observations.html.

Marin, D. (2014). Globalisation and the rise of the robots | VOX, CEPR policy portal. [online] *Voxeu.org*. Available at: https://voxeu.org/article/globalisation-and-rise-robots.

Market Research Future (Website) (n.d.). Robotic process automation for smartphone manufacturing market. [online] Available at: https://www.marketresearchfuture.com/reports/robotic-process-automation-for-smartphone-manufacturing-market-5219 [Accessed 28 Oct. 2020].

Marr, B. (2019). What is machine vision and how is it used in business today? [online] *Forbes*. Available at: https://www.forbes.com/sites/bernardmarr/2019/10/11/what-is-machine-vision-and-how-is-it-used-in-business-today/#6f0512f76939.

McBratney, A., McWhelan, B., Ancev, T., 2005. Future directions of precision agriculture. *Precision Agriculture*, 6, pp.7–23.

McKinsey Global Institute (2011). Internet matters: The net's sweeping impact on growth, jobs and prosperity. [online] *McKinsey*, pp.16–23. Available at: http://www.mckinsey.com/insights/mgi/research/technology_and_innovation/internet_matters.

McKinsey Global Institute (2021). Getting tangible about intangibles: The future of growth and productivity. [online] Available at: https://www.mckinsey.com/business-functions/marketing-and-sales/our-insights/getting-tangible-about-intangibles-the-future-of-growth-and-productivity# [Accessed November 1, 2021].

McWhelan, B.M., McBratney, A.B., 2003. Definition and interpretation of potential management zones in Australia. In Proceedings of the 11th Australian Agronomy Conference, Geelong, Victoria, February 2-6, 2003.

Milner and Rice, S. (2019). Robots are coming to a hospital near you. [online] *Fast Company*. Available at: https://www.fastcompany.com/90345453/robots-are-coming-to-a-hospital-near-you [Accessed October 12, 2021].

Mindy Support Website. How machine learning in automotive makes self-driving cars a reality. Available at: https://mindy-support.com/news-post/how-machine-learning-in-automotive-makes-self-driving-cars-a-reality/ [Accessed October 12, 2021].

Mint (Website) (2020). *Indian Online Food Delivery Market to Hit $8 bn by 2022: Report.* [online] Livemint. Available at: https://www.livemint.com/technology/tech-news/indian-online-food-delivery-market-to-hit-8-bn-by-2022-report-11580214173293.html [Accessed October 12, 2021].

Mitra, S. and Das, M. (2018). Thorny roses: A peep into the robotized economic future. [online] Available at: https://www.researchgate.net/publication/325398129_THORNY_ROSES_A_PEEP_INTO_THE_ROBOTISED_ECONOMIC_FUTURE [Accessed October 12, 2021].

Moses, L. (2017). The Washington Post's robot reporter has published 850 articles in the past year: Digiday. [online] *Digiday.* Available at: https://digiday.com/media/washington-posts-robot-reporter-published-500-articles-last-year/.

Narain, R. and Yadav, R. (1997). Impact of computerized automation on Indian manufacturing industries. *Technological Forecasting and Social Change*, 55(1), pp.83–98.

National Sample Survey Organization (1997). *Use of Durable Goods by Indian Households: 1993–94, Report No. 426.* [online] Available at: http://mospi.nic.in/sites/default/files/publication_reports/426_final.pdf.

National Sample Survey Organization (2012). Household consumption of various goods and services in India (July 2009–June 2010). Available at: http://mospi.nic.in/sites/default/files/publication_reports/nss_report_541.pdf.

Newberry, C. (2021) 24 YouTube Statistics that may surprise you: 2021 edition. *Hootsuite* (Website). Available at: https://blog.hootsuite.com/youtube-stats-marketers/ [Accessed October 10, 2021].

Niño-Zarazúa, M. and Addison, T. (2012). Redefining poverty in China and India: United Nations University. [online] *Unu.edu.* Available at: https://unu.edu/publications/articles/redefining-poverty-in-china-and-india.html.

Pantzar, M. (1992). The growth of product variety - a myth? *Journal of Consumer Studies and Home Economics*, 16(4), pp.345–362.

Pappas, S. (2020). Do face masks really reduce coronavirus spread? Experts have mixed answers. [online] *livescience.com.* Available at: https://www.livescience.com/are-face-masks-effective-reducing-coronavirus-spread.html.

Perez, S. (2021). US e-commerce on track for its first $1 trillion year by 2022, due to lasting pandemic impacts. [online] *TechCrunch.* Available at: https://techcrunch.com/2021/03/15/u-s-e-commerce-on-track-for-its-first-1-trillion-year-by-2022-due-to-lasting-pandemic-impacts/.

Planergy (Website). Artificial intelligence in procurement. [online]. Available at: https://planergy.com/blog/ai-in-procurement/

Rana, A. (September 21, 2018). Introduction: Reinforcement learning with OpenAI gym. Available at: https://towardsdatascience.com/reinforcement-learning-with-openai-d445c2c687d2 [Accessed October 12, 2021].

Research and Markets (July, 2021), *Autonomous/Driverless Car Market: Growth, Trends, COVID-19 Impact, and Forecasts (2021–2026), Report, ID: 4534508.* Available at: https://www.researchandmarkets.com/reports/4534508/autonomousdriverless-car-market-growth?utm_source=dynamic&utm_medium=GNOM&utm_code=5df483&utm_campaign=1365232+-+Global+Autonomous%2fDriverless+Car+Market+Projections%2c+2020-2025%3a+World+Market+Anticipating+a+CAGR+of+~18%25&utm_exec=joca220gnomd [Accessed October 12, 2021].

Rettner, R. (2020). Face masks may reduce COVID-19 spread by 85%, WHO-backed study suggests. [online] *livescience.com*. Available at: https://www.livescience.com/face-masks-eye-protection-covid-19-prevention.html [Accessed October 12, 2021].

Revfine (Website) (2019). 8 Examples of robots being used in the hospitality industry. [online] *Revfine.com*. Available at: https://www.revfine.com/robots-hospitality-industry/.

Rodrik, D. (2016). Premature deindustrialization. *Journal of Economic Growth*, 21(1), pp.1–33.

Roper, J.E. (n.d.) Revealed preference theory. *Britannica Website*. Available at: https://www.britannica.com/topic/revealed-preference-theory [Accessed October 10, 2021].

Rotman, D. (2020). We're not prepared for the end of Moore's Law. [online] *MIT Technology Review*. Available at: https://www.technologyreview.com/2020/02/24/905789/were-not-prepared-for-the-end-of-moores-law/.

Salmon, F. (2012). Udacity and the future of online universities. [online] *Reuters*. Available at: http://blogs.reuters.com/felix-salmon/2012/01/23/udacity-and-the-future-of-online-universities/.

Schaller, R.R. (1997). Moore's law: past, present and future. *IEEE Spectrum*, 34(6), pp.52–59.

Schaub, M. (2016). Is the future award-winning novelist a writing robot? [online] *Los Angeles Times*. Available at: http://www.latimes.com/books/jacketcopy/la-et-jc-novel-computer-writing-japan-20160322-story.html [Accessed 14 May 2021].

Selingo, J.J. (2013). *College (Un)bound: The Future of Higher Education and What it Means for Students*. New York: New Harvest.

Shubber, K. (2013). Artificial artists: when computers become creative. [online] *Wired UK*. Available at: http://www.wired.co.uk/article/can-computers-be-creative.

Shrivastav, Urvi (January 27, 2021). Digitisation in agriculture: A necessity for India. *BW BUSINESSWORLD*. Available at: http://www.businessworld.in/article/Digitisation-In-Agriculture-A-Necessity-For-India/27-01-2021-370573/

Sievo (Website). AI in procurement. [online] Available at: https://sievo.com/resources/ai-in-procurement [Accessed October 9, 2021].

Simon, H.A. (1955). A behavioural model of rational choice. *The Quarterly Journal of Economics*, 69(1), pp. 99–118.

Simon, H.A. (1956). Rational choice and the structure of the environment. *Psychological Review*, 63(2), pp. 129–13.

Simon, H.A. (1995). Artificial intelligence: an empirical science. *Artificial Intelligence*, [online] 77(1), pp.95–127. Available at: http://www.ic.unicamp.br/~wainer/cursos/2s2006/epistemico/simon-ia.pdf.

Singh, S. (2015). APMCs: The other side of the story. [online] *Business Line*. Available at: https://www.thehindubusinessline.com/opinion/apmcs-the-other-side-of-the-story/article6871346.ece [Accessed 14 May 2021].

Singh, S. (2019). The soon to be $200B online food delivery is rapidly changing the global food industry. [online] *Forbes*. Available at: https://www.forbes.com/sites/sarwantsingh/2019/09/09/the-soon-to-be-200b-online-food-delivery-is-rapidly-changing-the-global-food-industry/?sh=3e9b0d45b1bc.

Singularity Hub (2012). Automation reaches French Vineyards with a vine-pruning robot. [online] *Singularity Hub*. Available at: https://singularityhub.com/2012/11/26/automation-reaches-french-vineyards-with-a-vine-pruning-robot/.

Smith, S. (2013). Iamus: Is this the 21st century's answer to Mozart? *BBC News.* [online] 3 Jan. Available at: http://www.bbc.com/news/technology-20889644 [Accessed 14 May 2021].

Smith, A. (2016). *Public Predictions for the Future of Workforce Automation.* [online] Pew Research Center: Internet, Science & Tech. Available at: https://www.pewresearch .org/internet/2016/03/10/public-predictions-for-the-future-of-workforce -automation/.

Socialtables (2019). Robotics in hospitality: How will it impact guest experience? [online] *Social Tables.* Available at: https://www.socialtables.com/blog/hospitality -technology/robotics-experience/.

Spence, M. (2011). *The Next Convergence: The Future of Economic Growth in a Multispeed World*: New York: Farrar, Straus and Giroux, pp.225–237.

Srivastava, N. (2020). Rise of Robots and AI in the Coronavirus Era. [online] *Analytics India Magazine.* Available at: https://analyticsindiamag.com/rise-of-robots-and-ai-in -the-coronavirus-era/.

Statista (Website) (2020). *U.S.: Number of households 1960–2019.* [online] Statista. Available at: https://www.statista.com/statistics/183635/number-of-households-in -the-us/

Statista (Website) (2021). Netflix: number of subscribers worldwide 2021. Available at: https://www.statista.com/statistics/250934/quarterly-number-of-netflix-streaming -subscribers-worldwide/ [Accessed October 10, 2021].

Stéphane, R., Bortolon, C., Khoramshahi, M., Salesse, R. N., Burca, M., Marin, L., Bardy, B. G., Billard, A., Macioce, V. and Capdevielle, D. (2016). Humanoid robots versus humans: How is emotional valence of facial expressions recognized by individuals with schizophrenia? An exploratory study. *Schizophrenia Research*, 176(2–3), pp.506–513. Available at: https://doi.org/10.1016/j.schres.2016.06.001.

Tabuchi, H. (2010). For Sushi chain, conveyor belts carry profit. *The New York Times.* [online] 30 Dec. Available at: http://www.nytimes.com/2010/12/31/business/global /31sushi.html [Accessed 14 May 2021].

Taulli, T (June 5, 2020). Reinforcement learning: The next big thing for AI (artificial intelligence)?. *Forbes.* Available at: https://www.forbes.com/sites/tomtaulli/2020/06 /05/reinforcement-learning-the-next-big-thing-for-ai-artificial-intelligence/?sh =4e0acf3262ba [Accessed October 5, 2021].

The Associated Press (2019). Sewing robots set to bring jobs to Little Rock, fight foreign labor. [online] *Arkansas Online.* Available at: https://www.arkansasonline.com/news /2019/mar/26/sewing-robots-set-to-bring-jobs-to-lr-f/.

The Economist (2013). Coming home. [online] Available at: https://www.economist .com/sites/default/files/20130119_offshoring_davos.pdf.

Thomas, Z. (2020). Will Covid-19 accelerate the use of robots at work? [online] *BBC News* 19 Apr. Available at: https://www.bbc.com/news/technology-52340651.

Timmer, M. (2012). WIOD Working Paper No. 10. *WIOD: Content, Sources and Methods.* [online] Available at: http://www.wiod.org/publications/papers/wiod10.pdf.

United Nations (2019). World population prospects 2019. [online]. Available at: https:// population.un.org/wpp/Publications/Files/WPP2019_Highlights.pdf.

Vision Robotics Corporate (Website) (n.d.). PROJECTS. [online] *visionrobotics.* Available at: https://www.visionrobotics.com/projects [Accessed 12 Mar. 2017].

Whelan, B.M. and McBratney, A.B. (2003). Definition and interpretation of potential management zones in Australia. In: Proceedings of the 11th Australian Agronomy Conference, Geelong, Victoria, Feb. 2–6, 2003.

Wiegand, S. (2016). The impact of the television in 1950s America. [online] *Dummies*. Available at: https://www.dummies.com/education/history/american-history/the -impact-of-the-television-in-1950s-america/.

World Bank (2016). World development report, 2016: Digital dividends. *Choice Reviews Online*, 53(11).

World Bank (2020a). The World bank data. [online] *Worldbank.org*. Available at: https:// data.worldbank.org/indicator/NY.GDP.MKTP.KD.ZG?locations=PK.

World Bank (2020b). The World Bank in Ethiopia. [online] *World Bank*. Available at: https://www.worldbank.org/en/country/ethiopia/overview.

World Bank (2020c). The World Bank in Nigeria. [online] *World Bank*. Available at: https://www.worldbank.org/en/country/nigeria/overview.

World Economic Forum (2020). *The Future of Jobs Report 2020*. [online]. Available at: http://www3.weforum.org/docs/WEF_Future_of_Jobs_2020.pdf.

World Health Organization (2020). Q&A on coronaviruses (COVID-19). [online] *www .who.int*. Available at: https://www.who.int/emergencies/diseases/novel-coronavirus -2019/question-and-answers-hub/q-a-detail/q-a-coronaviruses.

Worldometer (Website) (n.d.). COVID-19 coronavirus pandemic. [online] Available at: https://www.worldometers.info/coronavirus/?utm_campaign=homeAdUOA?Si [Accessed 9 Nov. 2020].

Worldometer (Website) (2019). World population by year: Worldometers. [online] *Worldometers.info*. Available at: https://www.worldometers.info/world-population/ world-population-by-year/.

YellRobot (Website) (2018). Robots in hospitals are making deliveries and performing surgery. [online] *Robot News*. Available at: https://yellrobot.com/robots-in-hospitals/.

YouTube (Website). Available at: https://www.youtube.com/intl/ALL_in/howyoutube works/our-commitments/sharing-revenue/ [Accessed October 10, 2021].

YouTube Creator Academy (Website). Available at: https://creatoracademy.youtube.com /page/lesson/discovery#strategies-zippy-link-2 [Accessed October 10, 2021].

INDEX

Printed in the United States
by Baker & Taylor Publisher Services

Printed in the United States
by Baker & Taylor Publisher Services